Charles Hartwell Horne Cheyne, A. Freeman

An Alementary Treatise on the Planetary Theory

With a Collection of Problems

Charles Hartwell Horne Cheyne, A. Freeman

An Alementary Treatise on the Planetary Theory
With a Collection of Problems

ISBN/EAN: 9783337217921

Printed in Europe, USA, Canada, Australia, Japan

Cover: Foto ©berggeist007 / pixelio.de

More available books at **www.hansebooks.com**

AN

ELEMENTARY TREATISE

ON

THE PLANETARY THEORY,

WITH A COLLECTION OF PROBLEMS.

BY THE LATE

C. H. H. CHEYNE, M.A., F.R.A.S.

FORMERLY SCHOLAR OF ST JOHN'S COLLEGE, CAMBRIDGE,
AND ASSISTANT MATHEMATICAL MASTER IN WESTMINSTER SCHOOL.

THIRD EDITION.

EDITED BY THE

REV. A. FREEMAN, M.A., F.R.A.S.

RECTOR OF MURSTON, KENT,
AND LATE FELLOW OF ST JOHN'S COLLEGE, CAMBRIDGE.

London:

MACMILLAN AND CO.

1883

PREFACE TO THE THIRD EDITION.

AT the request of the Rev. Charles Cheyne, the father of the late C. H. H. Cheyne the author of this work, who to the great regret of all his relatives and friends died at Torquay on the 1st of January, 1877, I have undertaken the charge of this new edition. I have made but two changes in the body of the work, and have introduced but few notes. Inasmuch as the methods of Chapter II. are peculiarly the author's own I have not ventured to alter them : but I have instead added to the Appendix the Corollary at the end of Art. 5, a method of determining the variation of the elements (Art. 6) which is for the most part due to Pontécoulant with a slight addition of my own to make it complete, together with Art. 8 also largely due to the same author. I have also given some additional examples and have indicated the sources from which they were obtained.

If a new treatise on the Planetary Theory had to be written it would be necessary to digest the work of Jacobi, Hansen, Leverrier and Newcombe, to mention only a few eminent analysts. A brief account of a variety of methods might be found in Matthieu's *Dynamique*.

A biographical notice of the late author may be read in the *Monthly Notices* of the *Royal Astronomical Society*, Vol. XXXVII. pp. 147, 148. This I find as complete as is required in such a sketch of the too brief life of a diligent lover of Mathematical and Astronomical Science.

A. FREEMAN.

MURSTON RECTORY, KENT,
October 31st, 1883.

MEMOIRS BY C. H. H. CHEYNE.

1. On the Variation of the Elements in the Planetary Theory. (*Quarterly Journal of Pure and Applied Mathematics*, January, 1861.)

2. On the Variation of the Elements in the Planetary Theory. Second Paper. (*Quarterly Journal of Mathematics*, May, 1861.)

3. On the Variations of the Node and Inclination in the Planetary Theory. (*Quarterly Journal of Mathematics*, October, 1861.)

4. On the Equations of Motion of a Planet referred to Moving Axes. (*Oxford, Cambridge, and Dublin Messenger of Mathematics*, November, 1861.)

PREFACE TO THE FIRST EDITION.

IN this volume, an attempt has been made to produce a Treatise on the Planetary Theory, which, being elementary in character, should be so far complete, as to contain all that is usually required by students in this University. But it is not without diffidence that I submit my volume to their notice. In the earlier part of it, the methods which have been adopted are to some extent original*, and the general arrangement of the second Chapter will, it is believed, be found to be new. Through the kindness of the Publishers, a portion of Pratt's *Mechanical Philosophy* has been placed at my disposal. Of this I have availed myself, particularly in the Chapter on the Stability of the Planetary System; but, on the whole, comparatively little has been reprinted verbatim from that work. Among other sources of information, my obligations are mainly due to Pontécoulant's *Théorie Analytique du Système du Monde*, Airy's *Mathematical Tracts*, and Frost's *Planetary Theory* in the *Quarterly Journal of Mathematics*: but I have also referred to the

* Some of these have already appeared in Mathematical Journals.

Mécanique Céleste, the *Mécanique Analytique*, Mrs Somer-
ville's *Mechanism of the Heavens*, (a work forming a complete
Mathematical Treatise on Physical Astronomy,) a Memoir
by Prof. Donkin on the *Differential Equations of Dynamics*,
Phil. Trans. 1855, &c. A collection of Problems has been
added, taken chiefly from the Smith's Prize and Senate-
House Examination Papers of the last twenty years. In
conclusion, I would express my sincere thanks to Messrs.
A. Freeman, P. T. Main, and other friends, of St John's
College, for the valuable assistance which they have afforded
me, and would venture to hope that the work will be found
useful.

C. H. H. CHEYNE.

St John's College,
  October, 1862.

In the Second Edition comparatively few changes have
been made. The work has been revised, and, it is hoped,
in some degree improved. The Stability of the Planetary
System has been rather more fully treated, and an elegant
geometrical explanation of the formulæ for the secular
variations of the node and inclination introduced, for which
I am indebted to a paper by Mr H. M. Taylor, Fellow of
Trinity College, in the *Oxford, Cambridge and Dublin
Messenger of Mathematics.*

C. H. H. C.

1, Dean's Yard, Westminster,
  September, 1870.

CONTENTS.

CHAPTER I.

INTRODUCTION.

CHAPTER II.

FORMULÆ FOR CALCULATING THE ELEMENTS OF THE ORBIT.

CHAPTER III.

DEVELOPMENT OF THE DISTURBING FUNCTION.

CHAPTER IV.

SECULAR VARIATIONS OF THE ELEMENTS OF THE ORBIT. STABILITY OF THE PLANETARY SYSTEM.

CHAPTER V.

SECULAR VARIATIONS OF THE ELEMENTS CONTINUED. INTEGRATION OF THE DIFFERENTIAL EQUATIONS.

CHAPTER VI.

PERIODIC VARIATIONS OF THE ELEMENTS OF THE ORBIT.

CHAPTER VII.

DIRECT METHOD OF CALCULATING THE INEQUALITIES IN RADIUS
VECTOR, LONGITUDE, AND LATITUDE.

CHAPTER VIII.

ON THE EFFECTS WHICH A RESISTING MEDIUM WOULD PRODUCE IN
THE MOTIONS OF THE PLANETS.

APPENDIX.

THE PLANETARY THEORY.

CHAPTER I.

INTRODUCTION.

1. To determine the motion of a system of bodies subject only to their mutual attractions, is a problem the mathematical difficulties of which have not yet been overcome:
hence in the particular cases of this problem which Physical
Astronomy presents, recourse must be had to methods of approximation. Happily the arrangement of the Solar System
renders approximate methods possible, and in the skilful
hands of the Mathematicians of the last century, they have
been brought to a high state of perfection.

2. If the Sun were the only attracting body, the planets
would describe exact ellipses, agreeably to Kepler's first law ;
but in consequence of the attractions of the planets themselves, slight deviations from elliptic motion are produced.
The method of calculating these deviations, to which our
attention will chiefly be directed, is due to Euler, but was
subsequently extended and perfected by Lagrange : it consists in supposing the planets to move in ellipses, the elements (or arbitrary constants) of which are continually though
slowly changing*.

---

* The legitimacy of this hypothesis will appear when we come to treat
of the equations of motion. See Arts. 21 and 22.

C. P. T.   1

λ

3.   Now the elements of an elliptic orbit are (i) the *mean distance*, or semi-axis major, (ii) the *excentricity*, (iii) the *longitude of perihelion*, i.e. of the point of the orbit nearest to the Sun, (iv) the *longitude of the epoch**, or mean longitude at the epoch from which the time is reckoned, (v) the *inclination* of the plane in which the orbit lies to some fixed plane of reference, (vi) the *longitude of the ascending node*. Of these (i) and (ii) determine the magnitude of the orbit, (iii) determines its position in its own plane, (v) and (vi) determine the position of this plane, and (iv) has reference to the position of the body itself in its orbit.

If the planets moved accurately in ellipses, these would be constants : we must however be prepared to consider them as variable quantities, which it will be the object of the problem to determine.   They are termed the *elements of the orbit*.

4.   But further, not only is it found that the true orbit of a planet is not an ellipse, but that it is not even a plane curve, although the departure of the planet from the plane in which it is at any instant moving is extremely slow.   We define as the *plane of the orbit* the plane containing the radius vector and direction of motion of the planet at the instant under consideration.

5.   We shall suppose the Sun and planets so distant from each other that they may be considered to attract as if they were condensed into their respective centres of gravity; a supposition which would be rigorously true if these bodies were exactly spherical, and either of uniform density or composed of concentric spherical shells, the density of each shell being uniform throughout.   The errors, however, thus introduced into the motions of translation are found to be inappreciable for the planets, though not in the case of their satellites.   The motions of rotation will not be considered in the present treatise.

---

* Also briefly termed the *epoch*.

6. Moreover, since the masses of the planets are extremely small in comparison of that of the Sun, it follows that in cases where it is not necessary to carry the approximation beyond the first order of these masses, we are permitted to avail ourselves of the Principle of the Superposition of Small Motions, and thus to reduce the problem to a case of that of the Three Bodies.

7. So far the Theory of the Planets resembles that of the Moon, and the same method of treatment might be employed in both cases. But they differ in this respect: the ratio of the distances of the disturbed and disturbing bodies from the central one* is much smaller in the Lunar than in the Planetary Theory, so that if in the latter theory the approximation were made by means of series proceeding by powers of this ratio, it would be necessary to retain many more terms than are required in the former. On the other hand, the perturbations of the Moon are far larger than those of the planets, since in the former case the Sun, of which the mass is enormous, and the distance not proportionately great, is one of the disturbing bodies. For these reasons different methods of calculation are employed.

8. *To find an expression for the component in any direction of the force which disturbs the motion of a given planet relatively to the Sun.*

Let M denote the mass of the Sun, m, m', m'', &c., those of the planets, and suppose the relative motion of m required.

Let x, y, z, x', y', z', x'', y'', z'', &c., be the co-ordinates of m, m', m'', &c., referred to any system of rectangular axes

* By the *central body* is meant that whose attraction exercises the greatest influence on the body whose motion is required; the Sun, for instance, in the Theory of the Planets, and the Earth in that of the Moon. All the other attracting bodies are called *disturbing bodies*.

originating in the centre of gravity of the Sun; r, r', r'', &c., their distances from the origin; ρ', ρ'', &c., the distances of m', m'', &c., from m.

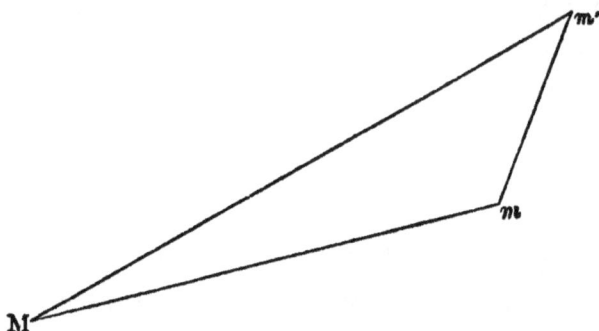

Now if to every body of the system we apply forces equal and opposite to those which act upon the Sun, we shall reduce the latter to rest without affecting the relative motion. Hence, considering the action of only one disturbing planet m', the forces acting upon m will be

$$\frac{M+m}{r^3} \text{ in direction } mM, \text{ or briefly, } \frac{\mu}{r^2},$$

$$\frac{m'}{\rho'^2} \text{ in direction } mm',$$

$$\frac{m'}{r'^2} \text{ in direction } m'M,$$

of which the last two constitute the disturbing force.

On the hypothesis of Art. 5, the components parallel to the axes of the forces acting on m, will be

$$-\frac{\mu x}{r^3} + \frac{m'(x'-x)}{\rho'^3} - \frac{m'x'}{r'^3},$$

$$-\frac{\mu y}{r^3} + \frac{m'(y'-y)}{\rho'^3} - \frac{m'y'}{r'^3},$$

$$-\frac{\mu z}{r^3} + \frac{m'(z'-z)}{\rho'^3} - \frac{m'z}{r'^3},$$

of which the terms containing m' are the disturbing forces.

Let s denote the length of the arc of any curve measured from some fixed point up to m: then the resolved part of the *disturbing force* parallel to the tangent at m to this curve will be

$$\frac{m'(x'-x)}{\rho'^3}\frac{dx}{ds} + \frac{m'(y'-y)}{\rho'^3}\frac{dy}{ds} + \frac{m'(z'-z)}{\rho'^3}\frac{dz}{ds}$$
$$-\left(\frac{m'x'}{r'^3}\frac{dx}{ds} + \frac{m'y'}{r'^3}\frac{dy}{ds} + \frac{m'z'}{r'^3}\frac{dz}{ds}\right),$$

or
$$\frac{d}{ds}\left\{\frac{m'}{\rho'} - \frac{m'}{r'^3}(xx'+yy'+zz')\right\},$$

on supposition that x, y, z, are alone affected by the process of differentiation, which may be written $\frac{dR'}{ds}$, if

$$R' = \frac{m'}{\rho'} - \frac{m'}{r'^3}(xx'+yy'+zz').$$

If we express in like manner the disturbing forces due to the action of m'', m''', &c., we shall have for the whole component in this direction

$$\frac{dR'}{ds} + \frac{dR''}{ds} + \ldots,$$

or $\frac{dR}{ds}$, if $R = R' + R'' + \ldots.$

The function R is called the *disturbing function*: the arbitrary curve employed above will be termed the *curve of reference.*

9. From the manner in which $\frac{dR}{ds}$ has been introduced, it appears that R is supposed to be expressed in terms of s and quantities which do not vary with s. It must, however, be borne in mind that in $\frac{dR}{ds}$ the variation is purely hypothetical, and has nothing whatever to do with the actual variation of R due to the motion of the planet.

For example, suppose the curve of reference a straight line parallel to the axis of x, and let R be expressed in terms of x, y and z; then in this case x only will vary, and the disturbing force parallel to the axis of x will be denoted by $\dfrac{dR}{dx}$, y and z being considered constant in the differentiation. Similarly, the disturbing forces parallel to the axes of y and z will be expressed by $\dfrac{dR}{dy}$ and $\dfrac{dR}{dz}$ respectively, the differential coefficients being strictly partial.

Again, suppose the curve of reference a circle with its plane parallel to that of xy, and its centre in the axis of z, and let R be expressed in terms of the polar co-ordinates (r_1, θ_1) of the projection of the planet on the plane of xy, and its distance (z) from this plane; then in this case θ_1 only will vary, and the disturbing force perpendicular to the projected radius vector will be expressed by $\dfrac{dR}{r_1 d\theta_1}$, r_1 and z being considered constant in the differentiation. Similarly, the forces parallel to the projected radius vector and to the axis of z, will be expressed by the partial differential coefficients $\dfrac{dR}{dr_1}$, $\dfrac{dR}{dz}$ respectively.

10. The disturbing function is independent of any particular system of co-ordinates that may be employed. For

$$R' = \frac{m'}{\rho'} - \frac{m'}{r'^3}(xx' + yy' + zz')$$

$$= \frac{m'}{\rho'} - \frac{m'r}{r'^2}\left(\frac{x}{r}\frac{x'}{r'} + \frac{y}{r}\frac{y'}{r'} + \frac{z}{r}\frac{z'}{r'}\right)$$

$$= \frac{m'}{\rho'} - \frac{m'r}{r'^2}\cos\omega,$$

if ω denote the inclination of r' to r.

11. *To express* R' *in terms of the polar co-ordinates of the projections of* m *and* m' *on a fixed plane, and of their distances from it.*

Take the fixed plane for that of xy : let r_1, r_1' be the projections of r, r' upon it, and θ_1, θ_1' the inclinations of r_1, r_1' to the axis of x; then

$$x = r_1 \cos \theta_1, \qquad y = r_1 \sin \theta_1,$$

$$x' = r_1' \cos \theta_1', \qquad y' = r_1' \sin \theta_1';$$

therefore $xx' + yy' + zz' = r_1 r_1' \cos (\theta_1 - \theta_1') + zz'$,

$$\rho'^2 = (x - x')^2 + (y - y')^2 + (z - z')^2$$

$$= r_1^2 + r_1'^2 - 2r_1 r_1' \cos (\theta_1 - \theta_1') + (z - z')^2,$$

$$r' = x'^2 + y'^2 + z'^2$$

$$= r_1'^2 + z'^2.$$

Hence by substitution,

$$R' = \frac{m'}{\{r_1^2 + r_1'^2 - 2r_1 r_1' \cos (\theta_1 - \theta_1') + (z - z')^2\}^{\frac{1}{2}}}$$
$$- \frac{m' \{r_1 r_1' \cos (\theta_1 - \theta_1') + zz'\}}{\{r_1'^2 + z'^2\}^{\frac{1}{2}}}.$$

12. In a subsequent Chapter we shall consider the development of R in terms of the time and the elements of the orbit, in a series ascending by powers and products of the eccentricities and inclinations, which for the principal planets are very small*. At present we shall content ourselves with shewing *how* R may be expressed in terms of these quantities.

* For the smaller planets and comets this is not the case; so that different methods of calculation are required for these bodies.

We shall assume that the equations connecting the co-ordinates, the time, and the elements in an elliptic orbit, hold also when the motion is disturbed.

13. *To explain how* R *may be expressed in terms of the time and the elements of the orbit.*

Let r, θ denote the radius vector and longitude of the disturbed planet, the latter being measured on a fixed plane of reference as far as the node, and thence on the plane of the orbit : let the elements be a the mean distance, e the excentricity, ϖ the longitude of perihelion, ϵ the longitude of the epoch, (the last two being measured in the same way as θ,) Ω the longitude of the node measured on the plane of reference, and i the inclination of the plane of the orbit to the plane of reference. Our object is to express R in terms of t and these elements.

Again, let θ_0, ϖ_0, ϵ_0, Ω_0 denote the longitudes of the planet, of perihelion, of the epoch, and of the node, measured entirely on the plane of the orbit.

Let a sphere be described with its centre coinciding with that of the Sun, and its radius of any magnitude : let the planes of reference and of the orbit cut it in the great circles NM, NP, then the lines of nodes will cut it in N; let the radius vector of the planet cut it in P, the projection of this radius on the plane of reference in M, and the lines from which θ, θ_0 are measured in L, O respectively. We shall suppose L to be the same origin as that from which θ_1 is measured in Art. 11.

Then in the figure $LM = \theta_1$, $LN + NP = \theta$, $OP = \theta_0$, $LN = \Omega$, the angle $PNM = i$, and $PM =$ the latitude of the planet which we shall denote by λ.

Hence from the right-angled triangle PNM,

$$\tan(\theta_1 - \Omega) = \cos i \tan(\theta - \Omega)\ldots\ldots\ldots(1),$$

$$\sin\lambda = \sin i \sin(\theta - \Omega)\ldots\ldots\ldots(2):$$

also

$$r_1 = r \cos\lambda\ldots\ldots\ldots\ldots\ldots(3),$$

$$z = r \sin\lambda\ldots\ldots\ldots\ldots\ldots(4).$$

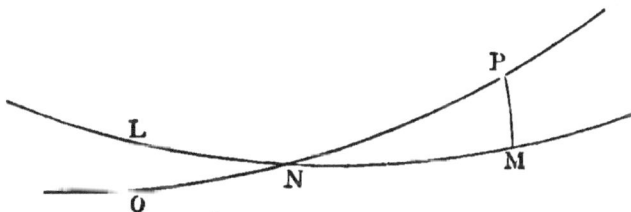

Again, from the formulæ of elliptic motion*,

$$r = a\{1 + \tfrac{1}{2}e^2 - e\cos(nt + \epsilon_0 - \varpi_0) - \tfrac{1}{2}e^2\cos 2(nt + \epsilon_0 - \varpi_0) - \ldots\},$$

$$\theta_0 = nt + \epsilon_0 + 2e\sin(nt + \epsilon_0 - \varpi_0) + \tfrac{5}{4}e^2\sin 2(nt + \epsilon_0 - \varpi_0) + \ldots\dagger:$$

but $\qquad \theta - \theta_0 = LN - ON = \epsilon - \epsilon_0 = \varpi - \varpi_0;$

therefore $\qquad \theta = \theta_0 + \epsilon - \epsilon_0, \quad \epsilon_0 - \varpi_0 = \epsilon - \varpi,$

and our formulæ become

$$r = a\{1 + \tfrac{1}{2}e^2 - e\cos(nt + \epsilon - \varpi) - \tfrac{1}{2}e^2\cos 2(nt + \epsilon - \varpi) - \ldots\}\ldots(5),$$

$$\theta = nt + \epsilon + 2e\sin(nt + \epsilon - \varpi) + \tfrac{5}{4}e^2\sin 2(nt + \epsilon - \varpi) + \ldots\ldots(6).$$

In Art. 11 we have expressed R' in terms of r_1, θ_1 and z; hence by equations (1) to (4) it may be expressed as a function of r, θ, Ω, and i: we may then substitute for r and θ from equations (5) and (6), and R' will be expressed in terms

* See Tait and Steele's *Dynamics*, Art. 162, *Fifth Edition.*

† n is termed the mean motion, and is connected with the mean distance by the equation $n^2a^3 = \mu$, where $\mu = M + m$.

of t and the elements of the orbit. Similarly R'', R''', &c., and therefore R may be expressed in terms of t and the elements.

14. We proceed to investigate certain relations which subsist between the partial differential coefficients of R with respect to the co-ordinates of the disturbed planet, and its partial differential coefficients with respect to the elements of the orbit. These will be useful in obtaining the formulæ by which the values of the elements are calculated.

We premise that when we speak of the partial differential coefficient of R with respect to one of the elements, we suppose R expressed in the manner indicated in the last Article, and that the time as well as the other elements are considered constant in the differentiation : when we speak of the partial differential coefficient of R with respect to r or θ, we suppose R expressed in terms of r, θ, i and Ω, which may be done by equations (1) to (4).

15. *To shew that* $\dfrac{dR}{d\theta} = \dfrac{dR}{d\epsilon} + \dfrac{dR}{d\varpi}$.

Equations (5) and (6) of Art. 13 may be written

$$r = f(nt + \epsilon - \varpi),$$
$$\theta - \varpi = \phi(nt + \epsilon - \varpi),$$

whence it follows that

$$\frac{dr}{d\epsilon} + \frac{dr}{d\varpi} = 0,$$

$$\frac{d\theta}{d\epsilon} + \frac{d\theta}{d\varpi} = 1.$$

Now since ϵ and ϖ enter into R only through r and θ,

$$\frac{dR}{d\epsilon} = \frac{dR}{dr}\frac{dr}{d\epsilon} + \frac{dR}{d\theta}\frac{d\theta}{d\epsilon},$$

$$\frac{dR}{d\varpi} = \frac{dR}{dr}\frac{dr}{d\varpi} + \frac{dR}{d\theta}\frac{d\theta}{d\varpi};$$

therefore, by addition,

$$\frac{dR}{d\epsilon} + \frac{dR}{d\varpi} = \frac{dR}{d\theta}.$$

16. *To shew that* $\dfrac{dR}{d\theta_1} = \dfrac{dR}{d\epsilon} + \dfrac{dR}{d\varpi} + \dfrac{dR}{d\Omega}.$

From equations (1) to (4) of Art. 13 we obtain

$$r_1 = \phi\,(r,\ \theta - \Omega,\ i),$$

$$\theta_1 - \Omega = \chi\,(\theta - \Omega,\ i),$$

$$z = \psi\,(r,\ \theta - \Omega,\ i),$$

where $\phi,\ \chi,\ \psi$ are symbols of functionality.

It follows that

$$\frac{dr_1}{d\theta} + \frac{dr_1}{d\Omega} = 0,$$

$$\frac{d\theta_1}{d\theta} + \frac{d\theta_1}{d\Omega} = 1,$$

$$\frac{dz}{d\theta} + \frac{dz}{d\Omega} = 0.$$

Now since by Art. 11, R is a function of r_1, θ_1, and z,

$$\frac{dR}{d\theta} = \frac{dR}{dr_1}\frac{dr_1}{d\theta} + \frac{dR}{d\theta_1}\frac{d\theta_1}{d\theta} + \frac{dR}{dz}\frac{dz}{d\theta},$$

$$\frac{dR}{d\Omega} = \frac{dR}{dr_1}\frac{dr_1}{d\Omega} + \frac{dR}{d\theta_1}\frac{d\theta_1}{d\Omega} + \frac{dR}{dz}\frac{dz}{d\Omega};$$

therefore, by addition,

$$\frac{dR}{d\theta} + \frac{dR}{d\Omega} = \frac{dR}{d\theta_1},$$

whence, by the last article,

$$\frac{dR}{d\theta_1} = \frac{dR}{d\epsilon} + \frac{dR}{d\varpi} + \frac{dR}{d\Omega}.$$

17. *To obtain* $\dfrac{dR}{de}$ *in terms of* $\dfrac{dR}{dr}$ *and* $\dfrac{dR}{d\theta}$.

If u denote the excentric anomaly, we have[*]

$$r = a\,(1 - e\,\cos u) \quad\dots\dots\dots\dots\dots\dots(1),$$

$$\tan\frac{\theta - \varpi}{2} = \sqrt{\left(\frac{1 + e}{1 - e}\right)}\tan\frac{u}{2} \quad\dots\dots\dots\dots(2),$$

$$nt + \epsilon - \varpi = u - e\,\sin u \dots\dots\dots\dots\dots\dots(3),$$

from which r and θ may be expressed in terms of t and the elements by eliminating u. Assuming r and θ so expressed, we proceed to obtain $\dfrac{dr}{de}$ and $\dfrac{d\theta}{de}$.

From (1), $\dfrac{dr}{de} = a\left(e\,\sin u\,\dfrac{du}{de} - \cos u\right),$

and from (3), $\dfrac{du}{de}\,(1 - e\,\cos u) - \sin u = 0\dots\dots\dots\dots(4)$;

eliminating $\dfrac{du}{de}$, we have

$$\frac{dr}{de} = a\left\{\frac{e\,\sin^2 u}{1 - e\,\cos u} - \cos u\right\} = a\left\{\frac{e - \cos u}{1 - e\,\cos u}\right\},$$

$$= a\,\frac{(1 + e)\,\sin^2\dfrac{u}{2} - (1 - e)\,\cos^2\dfrac{u}{2}}{(1 - e)\,\cos^2\dfrac{u}{2} + (1 + e)\,\sin^2\dfrac{u}{2}},$$

---

[*] See Tait and Steele's *Dynamics*, Arts. 160 and 161, *Fifth Edition.* This use of the auxiliary u was first suggested by Mr W. Pirie, *Camb. Math. Journal*, 1837, Vol. I., p. 47.

$$= a \frac{\tan^2 \frac{\theta - \varpi}{2} - 1}{1 + \tan^2 \frac{\theta - \varpi}{2}}, \text{ by (2),}$$

$$= - a \cos (\theta - \varpi).$$

Again, differentiating the logarithms of equation (2),

$$\frac{1}{\sin (\theta - \varpi)} \frac{d\theta}{de} = \frac{1}{2} \left(\frac{1}{1 + e} + \frac{1}{1 - e} \right) + \frac{1}{\sin u} \frac{du}{de};$$

eliminating $\frac{du}{de}$ by means of (4),

$$\frac{1}{\sin (\theta - \varpi)} \frac{d\theta}{de} = \frac{1}{1 - e^2} + \frac{1}{1 - e \cos u},$$

$$= a \left\{ \frac{1}{a (1 - e^2)} + \frac{1}{r} \right\}, \text{ by (1),}$$

$$= a \left\{ \frac{\mu}{h^2} + \frac{1}{r} \right\},$$

if $h^2 = \mu a (1 - e^2)$; therefore

$$\frac{d\theta}{de} = a \left(\frac{\mu}{h^2} + \frac{1}{r} \right) \sin (\theta - \varpi).$$

Now since R is a function of e only because it is a function of r and θ,

$$\frac{dR}{de} = \frac{dR}{dr} \frac{dr}{de} + \frac{dR}{d\theta} \frac{d\theta}{de},$$

$$= - a \cos (\theta - \varpi) \frac{dR}{dr} + a \left(\frac{\mu}{h^2} + \frac{1}{r} \right) \sin (\theta - \varpi) \frac{dR}{d\theta}.$$

Since $\theta - \varpi = \theta_0 - \varpi_0$, (see Art. 13,) this equation may be written

$$\frac{dR}{de} = - a \cos (\theta_0 - \varpi_0) \frac{dR}{dr} + a \left(\frac{\mu}{h^2} + \frac{1}{r} \right) \sin (\theta_0 - \varpi_0) \frac{dR}{d\theta},$$

under which form it will be useful in the next Chapter.

CHAPTER II.

18. WE now proceed to form equations of motion, taking the Sun's centre for the origin of co-ordinates, the radius vector of the planet for the axis of x, a perpendicular to it in the plane of the orbit for the axis of y, and a normal to this plane for the axis of z. With this system, it will be shewn that two of the resulting equations can be expressed in the same forms as if the planet moved in one plane.

Let x, y, z be the co-ordinates, u, v, w the velocities of the planet with reference to three rectangular axes originating in the Sun's centre, and moving with angular velocities ϕ_1, ϕ_2, ϕ_3 about their instantaneous positions : let X, Y, Z be the accelerations due to the impressed forces in the directions of the axes. Then (Routh's *Rigid Dynamics*, Arts. 244, 245 *Third Edition*),

$$\left. \begin{aligned} u &= \frac{dx}{dt} - y\phi_3 + z\phi_2 \\ v &= \frac{dy}{dt} - z\phi_1 + x\phi_3 \\ w &= \frac{dz}{dt} - x\phi_2 + y\phi_1 \end{aligned} \right\} \dots\dots\dots\dots\dots(1),$$

and the equations of motion are

$$X = \frac{du}{dt} - v\phi_3 + w\phi_2$$

$$Y = \frac{dv}{dt} - w\phi_1 + u\phi_3 \quad \Bigg\} \quad \ldots\ldots\ldots\ldots\ldots(2).$$

$$Z = \frac{dw}{dt} - u\phi_2 + v\phi_1$$

In these equations ϕ_1, ϕ_2, ϕ_3 are arbitrary; we propose so to determine them that the axis of x may coincide with the radius vector of the planet, and the plane of xy with the plane of the orbit.

In order that the axis of x may coincide with the radius vector of the planet, we must have

$$u = r, \quad y = 0, \quad z = 0,$$

always; and therefore

$$\frac{dx}{dt} = \frac{dr}{dt}, \quad \frac{dy}{dt} = 0, \quad \frac{dz}{dt} = 0 :$$

and in order that the plane of xy may coincide with the plane of the orbit, we must have

$$w = 0$$

always; and therefore

$$\frac{dw}{dt} = 0.$$

Hence equations (1) give

$$u = \frac{dr}{dt}, \quad v = r\phi_3, \quad \phi_2 = 0 ;$$

and equations (2) become

$$X = \frac{d^2 r}{dt^2} - r\phi_3^2,$$

$$Y = r\frac{d\phi_3}{dt} + 2\phi_3 \frac{dr}{dt},$$

$$Z = r\phi_3\phi_1.$$

Now let $\phi_3 = \dfrac{d\theta_0}{dt}$: thus for the equations of motion we have

$$X = \frac{d^2 r}{dt^2} - r \left(\frac{d\theta_0}{dt}\right)^2,$$

$$Y = \frac{1}{r} \frac{d}{dt}\left(r^2 \frac{d\theta_0}{dt}\right),$$

$$Z = r\phi_1 \frac{d\theta_0}{dt};$$

of which the first two are the same in form as if the plane of the orbit were at rest.

19. If we measure on this plane from the planet's radius vector in a direction contrary to that of motion an angle equal to θ_0, we arrive at what may be considered as the origin from which θ_0 is measured. Since this will be a point having no angular velocity about the axis of z, which is normal to the plane of the orbit, it is said to be *fixed in the plane of the orbit**.

20. We shall for the present confine our attention to the first two of the above equations.

In order to find the components of the disturbing force parallel and perpendicular to the radius vector of the planet, let us take first the radius vector as the curve of reference; then $s = r$, and R being supposed expressed as a function of r, θ, Ω, and i (see Art. 13), we have

$$\frac{dR}{ds} = \frac{dR}{dr}\frac{dr}{ds} + \frac{dR}{d\theta}\frac{d\theta}{ds} + \frac{dR}{d\Omega}\frac{d\Omega}{ds} + \frac{dR}{di}\frac{di}{ds}$$

$$= \frac{dR}{dr},$$

---

* It is necessary to define the expression *fixed in the plane of the orbit*, since the definition of Art. 4 is not sufficient completely to regulate the motion of this plane, though affording it a distinct geometrical position.

since θ, Ω and i do not vary with s. Hence the disturbing force in direction of the radius vector $= \dfrac{dR}{dr}$.

Again, let us take as the curve of reference a circle in the plane of the orbit, with its centre coinciding with that of the Sun; then $\delta s = r \delta \theta$, and we have

$$\frac{dR}{ds} = \frac{1}{r}\frac{dR}{d\theta},$$

since r, Ω, and i do not vary with s. Hence the disturbing force perpendicular to the radius vector $= \dfrac{1}{r}\dfrac{dR}{d\theta}$.

We have then

$$X = -\frac{\mu}{r^2} + \frac{dR}{dr}, \quad Y = \frac{1}{r}\frac{dR}{d\theta},$$

and the equations become

$$\frac{d^2r}{dt^2} - r\left(\frac{d\theta_0}{dt}\right)^2 = -\frac{\mu}{r^2} + \frac{dR}{dr} \dots \dots \dots \dots \dots (1),$$

$$\frac{d}{dt}\left(r^2\frac{d\theta_0}{dt}\right) = \frac{dR}{d\theta} \dots \dots \dots \dots \dots \dots (2).$$

21. These equations do not admit of rigorous integration, but we may reduce them by the method of the Variation of Parameters to a system of differential equations of the first order. The principle of this method may be explained as follows. Suppose it required to integrate the equations

$$\left.\begin{array}{l} \phi_1\left(x, y, t, \dfrac{dx}{dt}, \dfrac{dy}{dt}, \dfrac{d^2x}{dt^2}, \dfrac{d^2y}{dt^2}\right) = P_1 \\[3mm] \phi_2\left(x, y, t, \dfrac{dx}{dt}, \dfrac{dy}{dt}, \dfrac{d^2x}{dt^2}, \dfrac{d^2y}{dt^2}\right) = P_2 \end{array}\right\} \dots \dots (i),$$

where P_1, P_2 are functions of t. The solution of these equations can be made to depend upon that of the equations

$\phi_1 = 0$, $\phi_2 = 0$. Suppose the four first integrals of these last equations to be

$$\left. \begin{array}{l} \chi_1\left(x, y, t, \dfrac{dx}{dt}, \dfrac{dy}{dt}\right) = c_1 \\[2mm] \chi_2\left(x, y, t, \dfrac{dx}{dt}, \dfrac{dy}{dt}\right) = c_2 \\[2mm] \chi_3\left(x, y, t, \dfrac{dx}{dt}, \dfrac{dy}{dt}\right) = c_3 \\[2mm] \chi_4\left(x, y, t, \dfrac{dx}{dt}, \dfrac{dy}{dt}\right) = c_4 \end{array} \right\} \quad \ldots\ldots\ldots\ldots\text{(ii)},$$

where c_1, c_2, c_3, c_4 are arbitrary constants or parameters. The method of the *Variation of Parameters* consists in so determining c_1, c_2, c_3 and c_4 as functions of t, that these integrals $\Big($and therefore the two final integrals of the equations $\phi_1 = 0$, $\phi_2 = 0$, which can be obtained from equations (ii) by eliminating $\dfrac{dx}{dt}$ and $\dfrac{dy}{dt}\Big)$ shall satisfy equations (i). That c_1, c_2, c_3, and c_4 *can* be so determined, may be seen as follows: by the solution of equations (i), values of x and y and therefore of $\dfrac{dx}{dt}$ and $\dfrac{dy}{dt}$ can be found as functions of t and constant quantities; if these be substituted in equations (ii) the requisite values of c_1, c_2, c_3 and c_4 will be obtained. For an example of the application of this method, see Boole's *Differential Equations*, Chap. IX. Art. 11.

22. If in equations (1) and (2) of Art. 20 we put $R = 0$, and then integrate them, we obtain

$$\frac{1}{r} = \frac{\mu}{h^2}\{1 + e\cos(\theta_0 - \varpi_0)\} \quad \ldots\ldots\ldots\ldots\text{(3)},$$

$$\frac{dr}{dt} = \frac{\mu e}{h}\sin(\theta_0 - \varpi_0) \quad \ldots\ldots\ldots\ldots\text{(4)},$$

$$r^2 \frac{d\theta_0}{dt} = h \quad \dots\dots\dots\dots\dots\dots\dots(5),$$

where h, e, ϖ_0 are the constants of integration. Equation (3) indicates motion in an ellipse, of which e is the excentricity, ϖ_0 the longitude of perihelion, and h twice the area described in an unit of time. If the mean distance in this ellipse be denoted by a, we have in addition

$$h^2 = \mu a \, (1 - e^2) \quad \dots\dots\dots\dots\dots(6).$$

We shall assume (in accordance with the principles of the method of the Variation of Parameters) the first and second integrals of equations (1) and (2), together with equation (6), to retain the same forms when R is not zero; h, e, ϖ_0, and a being in this case considered variable*.

The values of these elements are to be obtained from the condition that the above integrals shall satisfy equations (1) and (2).

23. If their values as calculated for any given time be substituted in equation (3), it will represent an ellipse having a contact of the first order with the actual orbit, since the values of $\dfrac{dr}{dt}$ and $\dfrac{d\theta_0}{dt}$ at the common point will be the same for both curves. It is termed the *instantaneous ellipse*, since the planet may for an infinitely small time be supposed to move in it. Moreover, the velocity and direction of motion of the planet will be the same as if it moved in this ellipse, so that if at any time the disturbing force were to cease, the planet would continue to move in the instantaneous ellipse constructed for that time. This is accordingly sometimes given as the definition of the instantaneous ellipse.

* We shall also for convenience suppose the equation $n^2 a^3 = \mu$ to hold in the disturbed orbit, n being of course considered variable.

24. *To obtain formulæ for calculating the elements of the instantaneous ellipse at any time.*

Suppose the value of c required, where c denotes any one of the elements. From equations (3), (4), and (6) we may, by eliminating the other elements, obtain c as a function of r, θ_0, $\dfrac{dr}{dt}$, and h^*: let then

$$c = f(r, \theta_0, r', h),$$

where r' is written for $\dfrac{dr}{dt}$. Differentiating, we have

$$\frac{dc}{dt} = \frac{df}{dr}\frac{dr}{dt} + \frac{df}{d\theta_0}\frac{d\theta_0}{dt} + \frac{df}{dr'}\frac{dr'}{dt} + \frac{df}{dh}\frac{dh}{dt}.$$

Now $\dfrac{dr'}{dt} = \dfrac{d^2r}{dt^2} = r\left(\dfrac{d\theta_0}{dt}\right)^2 - \dfrac{\mu}{r^2} + \dfrac{dR}{dr}$, from equation (1);

$$\frac{dh}{dt} = \frac{d}{dt}\left(r^2\frac{d\theta_0}{dt}\right) = \frac{dR}{d\theta}, \text{ from equation (2);}$$

therefore $\dfrac{dc}{dt} = \dfrac{df}{dr}\dfrac{dr}{dt} + \dfrac{df}{d\theta_0}\dfrac{d\theta_0}{dt} + \dfrac{df}{dr'}\left\{r\left(\dfrac{d\theta_0}{dt}\right)^2 - \dfrac{\mu}{r^2}\right\}$

$$+ \frac{df}{dr'}\frac{dR}{dr} + \frac{df}{dh}\frac{dR}{d\theta}.$$

But, since by hypothesis, if R were zero and c constant, our assumed integrals would still satisfy the differential equations, we have (making R zero and c constant)

$$0 = \frac{df}{dr}\frac{dr}{dt} + \frac{df}{d\theta_0}\frac{d\theta_0}{dt} + \frac{df}{dr'}\left\{r\left(\frac{d\theta_0}{dt}\right)^2 - \frac{\mu}{r^2}\right\};$$

therefore $\dfrac{dc}{dt} = \dfrac{df}{dr'}\dfrac{dR}{dr} + \dfrac{df}{dh}\dfrac{dR}{d\theta}.$

---

* We retain h for convenience, in preference to replacing it by $r^2\dfrac{d\theta_0}{dt}$.

Hence in obtaining the formulæ for calculating the elements of the orbit, we may proceed as follows. From equations (3), (4), and (6) we may express the element required as a function of r, θ_0, $\dfrac{dr}{dt}$, and h. We may then, by differentiating the resulting equation with respect to t as if r and θ_0 were constants, writing $\dfrac{dR}{dr}$ for $\dfrac{d^2r}{dt^2}$, and $\dfrac{dR}{d\theta}$ for $\dfrac{dh}{dt}$, and eliminating, if necessary, $\dfrac{dr}{dt}$ and h by means of equations (4) and (6), obtain the differential coefficient of the element required in terms of the elements, the co-ordinates of the planet, and the disturbing force. The result, however, will in every case admit of being expressed in terms of the elements and of the differential coefficients of R with respect to them, a form under which it is very convenient of application.

25. *To obtain a formula for calculating the mean distance.*

From equations (3), (4) and (6), if e and ϖ_0 be eliminated, we shall find

$$-\frac{\mu}{a} = \left(\frac{dr}{dt}\right)^2 + \frac{h^2}{r^2} - \frac{2\mu}{r}.$$

Differentiating as if r were constant, and writing $\dfrac{dR}{dr}$ for $\dfrac{d^2r}{dt^2}$, $\dfrac{dR}{d\theta}$ for $\dfrac{dh}{dt}$, we have

$$\frac{\mu}{a^2}\frac{da}{dt} = 2\left(\frac{dR}{dr}\frac{dr}{dt} + \frac{dR}{d\theta}\frac{d\theta_0}{dt}\right).$$

Now since $\qquad r = f(nt + \epsilon - \varpi)$,

$$\theta_0 - \varpi_0 = \theta - \varpi = \phi(nt + \epsilon - \varpi),$$

and the forms of $\dfrac{dr}{dt}$ and $\dfrac{d\theta_0}{dt}$ are the same as if the elements were constant, we have

$$\frac{dr}{dt} = nf'(nt + \epsilon - \varpi) = n\frac{dr}{d\epsilon},$$

and similarly,
$$\frac{d\theta_0}{dt} = n\frac{d\theta_0}{d\epsilon} = n\frac{d\theta}{d\epsilon};$$

therefore
$$\frac{\mu}{a^2}\frac{da}{dt} = 2n\left(\frac{dR}{dr}\frac{dr}{d\epsilon} + \frac{dR}{d\theta}\frac{d\theta}{d\epsilon}\right)$$
$$= 2n\frac{dR}{d\epsilon},$$

or
$$\frac{da}{dt} = \frac{2na^2}{\mu}\frac{dR}{d\epsilon}.$$

26. This formula may also be obtained as follows. If s denote an arc of the actual path of the planet measured from some fixed point to its position at time t, we have the equation of motion
$$\frac{d^2s}{dt^2} = -\frac{\mu}{r^2}\frac{dr}{ds} + \frac{dR}{ds},$$

and by a known formula
$$\left(\frac{ds}{dt}\right)^2 = \frac{2\mu}{r} - \frac{\mu}{a}.$$

Differentiating the latter we obtain
$$2\frac{ds}{dt}\frac{d^2s}{dt^2} = -\frac{2\mu}{r^2}\frac{dr}{dt} + \frac{\mu}{a^2}\frac{da}{dt},$$

and, multiplying the former by $2\frac{ds}{dt}$,
$$2\frac{ds}{dt}\frac{d^2s}{dt^2} = -\frac{2\mu}{r^2}\frac{dr}{dt} + 2\frac{dR}{ds}\frac{ds}{dt};$$

therefore
$$\frac{\mu}{a^2}\frac{da}{dt} = 2\frac{dR}{ds}\frac{ds}{dt}$$
$$= 2\frac{d(R)}{dt},$$

where $\frac{d(R)}{dt}$ denotes the differential coefficient of R with respect to t, only so far as R involves t through involving the co-ordinates and elements of the *disturbed* planet. Now since the velocity of the planet at any time can be expressed in terms of the co-ordinates and elements of the instantaneous ellipse constructed for that time in the same form as if it moved in this ellipse, its component in any direction can also be so expressed. Hence, considering R as a function of r_1, θ_1, z (see Art. 11), the values of $\dfrac{dr_1}{dt}$, $\dfrac{d\theta_1}{dt}$, $\dfrac{dz}{dt}$, and therefore of $\dfrac{d\,(R)}{dt}$, may be expressed in the same forms as if the elements were invariable. Since, then, t always occurs in R coupled with ϵ in the expression $nt + \epsilon$, we have

$$\frac{d\,(R)}{dt} = n\frac{dR}{d\,(nt+\epsilon)} = n\frac{dR}{d\epsilon};$$

therefore

$$\frac{da}{dt} = \frac{2na^2}{\mu}\frac{dR}{d\epsilon}.$$

27. *To obtain a formula for calculating the excentricity.*

From equations (3) and (4), if ϖ_0 be eliminated, we obtain

$$\left(\frac{dr}{dt}\right)^2 = \frac{\mu^2 e^2}{h^2} - \left(\frac{h}{r} - \frac{\mu}{h}\right)^2.$$

Differentiating as if $\dot{r}$ were constant, and writing $\dfrac{dR}{dr}$ for $\dfrac{d^2r}{dt^2}$,

$$\frac{dR}{dr}\frac{dr}{dt} = \frac{\mu^2 e}{h^2}\frac{de}{dt} - \left\{\frac{\mu^2 e^2}{h^3} + \left(\frac{1}{r}+\frac{\mu}{h^2}\right)\left(\frac{h}{r}-\frac{\mu}{h}\right)\right\}\frac{dh}{dt}$$

$$= \frac{\mu^2 e}{h^2}\frac{de}{dt} + \frac{\mu^2(1-e^2)}{h^3}\frac{dR}{d\theta} - \frac{h}{r^2}\frac{dR}{d\theta},$$

since

$$\frac{dh}{dt} = \frac{dR}{d\theta};$$

therefore $\dfrac{\mu^2 e}{h^2}\dfrac{de}{dt} = \dfrac{dR}{dr}\dfrac{dr}{dt} + \dfrac{dR}{d\theta}\dfrac{d\theta_0}{dt} - \dfrac{\mu^2(1-e^2)}{h^3}\dfrac{dR}{d\theta}$

$$= n\left(\dfrac{dR}{dr}\dfrac{dr}{d\epsilon} + \dfrac{dR}{d\theta}\dfrac{d\theta}{d\epsilon}\right) - \dfrac{\mu^2(1-e^2)}{h^3}\dfrac{dR}{d\theta},$$

therefore $\dfrac{de}{dt} = \dfrac{nh^2}{\mu^2 e}\dfrac{dR}{d\epsilon} - \dfrac{1-e^2}{he}\left(\dfrac{dR}{d\epsilon} + \dfrac{dR}{d\varpi}\right),$ (Art. 15),

$$= \dfrac{na(1-e^2)}{\mu e}\dfrac{dR}{d\epsilon} - \dfrac{na\sqrt{(1-e^2)}}{\mu e}\left(\dfrac{dR}{d\epsilon} + \dfrac{dR}{d\varpi}\right),$$

since $\quad h^2 = \mu a(1-e^2),$ and $n^2 a^3 = \mu.$

28.   This formula may also be deduced from that of the mean distance, by means of the equation

$$h^2 = \mu a(1-e^2).$$

We have $2h\dfrac{dh}{dt} = \mu(1-e^2)\dfrac{da}{dt} - 2\mu ae\dfrac{de}{dt},$

or $2na^2\sqrt{(1-e^2)}\dfrac{dR}{d\theta} = 2na^2(1-e^2)\dfrac{dR}{d\epsilon} - 2\mu ae\dfrac{de}{dt};$

therefore $\dfrac{de}{dt} = \dfrac{na(1-e^2)}{\mu e}\dfrac{dR}{d\epsilon} - \dfrac{na\sqrt{(1-e^2)}}{\mu e}\left(\dfrac{dR}{d\epsilon} + \dfrac{dR}{d\varpi}\right).$

29.   *To obtain a formula for calculating the longitude of perihelion.*

From equations (3) and (4), if e be eliminated, we obtain

$$\dfrac{dr}{dt}\cot(\theta_0 - \varpi_0) = \dfrac{h}{r} - \dfrac{\mu}{h}.$$

Differentiating as if r and θ_0 were constant, and writing $\dfrac{dR}{dr}$ for $\dfrac{d^2 r}{dt^2}$, we obtain

$$\dfrac{dr}{dt}\csc^2(\theta_0 - \varpi_0)\dfrac{d\varpi_0}{dt} + \cot(\theta_0 - \varpi_0)\dfrac{dR}{dr} = \left(\dfrac{1}{r} + \dfrac{\mu}{h^2}\right)\dfrac{dh}{dt}$$

$$= \left(\dfrac{1}{r} + \dfrac{\mu}{h^2}\right)\dfrac{dR}{d\theta};$$

therefore $\dfrac{dr}{dt} \operatorname{cosec} (\theta_0 - \pi_0) \dfrac{d\varpi_0}{dt}$

$$= - \cos (\theta_0 - \varpi_0) \dfrac{dR}{dr} + \left(\dfrac{1}{r} + \dfrac{\mu}{h^2}\right) \sin (\theta_0 - \varpi_0) \dfrac{dR}{d\theta}$$

$$= \dfrac{1}{a}\dfrac{dR}{de} \quad \text{(Art. 17)};$$

but from equation (4) $\dfrac{dr}{dt} \operatorname{cosec} (\theta_0 - \varpi_0) = \dfrac{\mu\theta}{h}$,

therefore $\dfrac{d\varpi_0}{dt} = \dfrac{h}{\mu e a}\dfrac{dR}{de} = \dfrac{na \sqrt{(1-e^2)}}{\mu e}\dfrac{dR}{de}.$

Now if ϖ denote the longitude of perihelion measured on the plane of reference as far as the node, and thence on the plane of the orbit, Ω the longitude of the node on the plane of reference, Ω_0 its longitude on that of the orbit, we have

$$\varpi - \varpi_0 = \Omega - \Omega_0;$$

therefore $\dfrac{d\varpi}{dt} = \dfrac{d\varpi_0}{dt} + \dfrac{d\Omega}{dt} - \dfrac{d\Omega_0}{dt}.$

Now $\dfrac{d\Omega}{dt}$ is the angular velocity of the line of nodes on the plane of reference, $\dfrac{d\Omega_0}{dt}$ its angular velocity on the plane of the orbit; and since the plane of reference is fixed, the former is the total angular velocity: hence

$$\dfrac{d\Omega_0}{dt} = \dfrac{d\Omega}{dt} \cos i;$$

therefore $\dfrac{d\varpi}{dt} = \dfrac{d\varpi_0}{dt} + (1 - \cos i) \dfrac{d\Omega}{dt};$

or, substituting for $\dfrac{d\Omega}{dt}$ from Art. 31 or 35,

$$\dfrac{d\varpi}{dt} = \dfrac{na \sqrt{(1-e^2)}}{\mu e}\dfrac{dR}{de} + \dfrac{na \tan \dfrac{i}{2}}{\mu \sqrt{(1-e^2)}}\dfrac{dR}{di}.$$

To obtain formulæ for calculating the longitude of the node, and the inclination.

30. We now return to our third equation of motion,

$$r\phi_1 \frac{d\theta_o}{dt} = Z,$$

or, as it may be written,

$$h\phi_1 = Zr \dots\dots\dots\dots\dots(1).$$

We have seen (Art. 18), that $\phi_2 = 0$; hence the motion of the plane of xy, which coincides with the plane of the orbit, is compounded of the angular velocities ϕ_1 about the axis of x, and ϕ_3 about the axis of z. Now the former is equivalent to an angular velocity $\phi_1 \cos(\theta - \Omega)$ about the line of nodes, and an angular velocity $\phi_1 \sin(\theta - \Omega)$ about an axis perpendicular to it in the plane of the orbit: but the angular velocities of the plane of the orbit about these axes are $\dfrac{di}{dt}$ and $\sin i \cdot \dfrac{d\Omega}{dt}$ respectively; therefore

$$\phi_1 \cos(\theta - \Omega) = \frac{di}{dt}, \quad \phi_1 \sin(\theta - \Omega) = \sin i \cdot \frac{d\Omega}{dt}.$$

Hence, by equation (1),

$$h\frac{di}{dt} = Zr \cos(\theta - \Omega) \dots\dots\dots\dots(2),$$

$$h \sin i \frac{d\Omega}{dt} = Zr \sin(\theta - \Omega) \dots\dots\dots\dots(3).$$

31. In order to determine Z we must suppose the curve of reference perpendicular to the plane of the orbit. If we denote by s an arc of this curve measured from some fixed point up to the planet, we have by Art. 8, $Z = \dfrac{dR}{ds}$. Now in the same way that the position of the planet is known when we know r, θ, i and Ω, the position of any point on

the curve of reference may be determined by its polar co-
ordinates r, θ on a plane passing through it and the Sun, the
inclination i of this plane to the plane of reference, and the
longitude Ω of its node.   Since, however, an infinite number
of planes can be drawn through two given points, we must
introduce some further condition to fix the position of that
on which r and θ are measured*.

Different forms of expression will be obtained for Z
according as different conditions are assigned.   First suppose the plane to pass through SN, the planet's line of
nodes: let P be the position of the planet, draw PC perpendicular to SN, and take for the curve of reference a
circle AP with centre C.   Through SN draw a plane inclined at a small angle δi to the plane of the orbit, cutting
the circle AP in p, and let $Pp = \delta s$; then since Pp is perpendicular to the plane SNP,

$$Pp = CP \cdot \delta i,$$

or

$$\delta s = r \sin (\theta - \Omega) \, \delta i;$$

therefore

$$\frac{di}{ds} = \frac{1}{r \sin (\theta - \Omega)}.$$

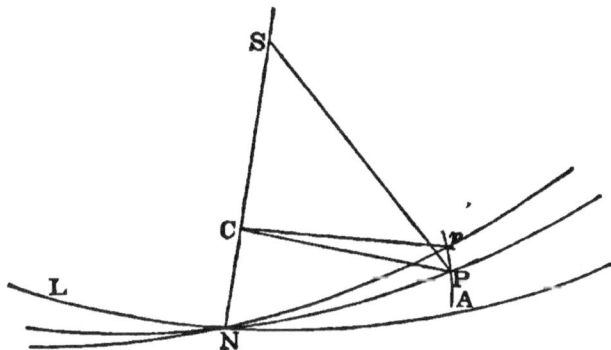

Now, (see Art. 13), R may be expressed in the form

$$R = f(r, \theta, \Omega, i),$$

in which, if the series for r and θ be substituted, R will be expressed as a function of t and the elements. Hence

$$\frac{dR}{ds} = \frac{dR}{dr}\frac{dr}{ds} + \frac{dR}{d\theta}\frac{d\theta}{ds} + \frac{dR}{d\Omega}\frac{d\Omega}{ds} + \frac{dR}{di}\frac{di}{ds},$$

and

$$\frac{dr}{ds} = 0, \quad \frac{d\theta}{ds} = 0, \quad \frac{d\Omega}{ds} = 0, \quad \frac{di}{ds} = \frac{1}{r\sin(\theta - \Omega)};$$

therefore

$$\frac{dR}{ds} = \frac{1}{r\sin(\theta - \Omega)}\frac{dR}{di}.$$

On substituting this value for Z in equation (3), we obtain

$$h\sin i \frac{d\Omega}{dt} = \frac{dR}{di},$$

or since $h^2 = \mu a(1 - e^2)$, and $n^2 a^3 = \mu$,

$$\frac{d\Omega}{dt} = \frac{na}{\mu\sqrt{(1 - e^2)}\sin i}\frac{dR}{di}.$$

32. By substituting the value of Z found in the preceding Article in equation (2) of Art. 30, we might, of course, obtain an expression for $\frac{di}{dt}$, but this would involve θ, and consequently be in a form inconvenient for calculation. We proceed, then, to obtain an expression for Z by means of which θ may be eliminated.

33. Suppose the plane on which r and θ are measured, instead of passing through SN, to pass through a line SC in the plane of the orbit perpendicular to SN: let P be the position of the planet, draw PC perpendicular to SC, and take for the curve of reference a circle AP with centre C.

Through SC draw a plane inclined at a small angle to the plane of the orbit, cutting the sphere in the great circle nmp, and the circle AP in p. Draw Nm perpendicular to np.

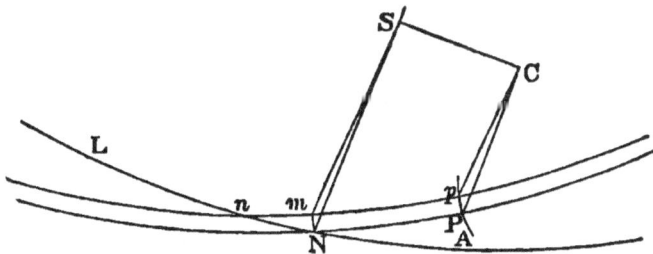

Then Nm which measures the inclination of the two planes $= - \delta\Omega \sin i$, and

$$CP = r \cos (\theta - \Omega);$$

hence if $Pp = \delta s$,

$$\delta s = - r \cos (\theta - \Omega) \, \delta\Omega \sin i;$$

therefore

$$\frac{\delta\Omega}{ds} = \frac{-1}{r \cos (\theta - \Omega) \sin i}.$$

Again

$$\delta\theta = Ln + np - (LN + NP)$$

$$= nm - nN$$

$$= - \delta\Omega \cos i + \delta\Omega;$$

therefore

$$\frac{d\theta}{ds} = (1 - \cos i) \frac{d\Omega}{ds};$$

also

$$\frac{dr}{ds} = 0, \; \frac{di}{ds} = 0.$$

Hence

$$\frac{dR}{ds} = \frac{dR}{d\theta}\frac{d\theta}{ds} + \frac{dR}{d\Omega}\frac{d\Omega}{ds}$$

$$= - \left\{ \frac{dR}{d\Omega} + (1 - \cos i) \frac{dR}{d\theta} \right\} \frac{1}{r \cos (\theta - \Omega) \sin i}.$$

On substituting this value for Z in equation (2), we have

$$\frac{di}{dt} = -\frac{1}{h \sin i}\left\{\frac{dR}{d\Omega} + (1 - \cos i)\,\frac{dR}{d\theta}\right\}$$

$$= -\frac{1}{h \sin i}\left\{\frac{dR}{d\Omega} + 2\sin^2\frac{i}{2}\left(\frac{dR}{d\epsilon} + \frac{dR}{d\varpi}\right)\right\} \quad \text{(Art. 15)},$$

$$= -\frac{na}{\mu\sqrt{(1-e^2)}}\left\{\frac{1}{\sin i}\frac{dR}{d\Omega} + \tan\frac{i}{2}\left(\frac{dR}{d\epsilon} + \frac{dR}{d\varpi}\right)\right\}.$$

The only remaining element is the epoch, but before proceeding to obtain a formula for its calculation, we shall give another method of obtaining the results of Arts. 31 and 33.

34. *To obtain a formula for calculating the inclination.* (*Second method.*)

If the motion of the planet be referred to the polar coordinates of its projection on the fixed plane of reference and its distance from this plane, we have the equation

$$\frac{1}{r_1}\frac{d}{dt}\left(r_1^2\frac{d\theta_1}{dt}\right) = \frac{1}{r_1}\frac{dR}{d\theta_1} \quad \text{(See Art. 9)},$$

or

$$\frac{d}{dt}\left(r_1^2\frac{d\theta_1}{dt}\right) = \frac{dR}{d\theta_1}.$$

Now if δA, δA_1 denote the vectorial areas swept out in the time δt on the plane of the orbit and the plane of reference respectively, we have

$$\delta A_1 = \delta A \cos i:$$

but

$$\delta A_1 = \frac{1}{2}r_1^2\delta\theta_1, \quad \delta A = \frac{1}{2}r^2\delta\theta_0;$$

therefore

$$r_1^2\frac{d\theta_1}{dt} = r^2\frac{d\theta_0}{dt}\cos i = h\cos i.$$

Hence our equation of motion becomes

$$\frac{d}{dt}(h \cos i) = \frac{dR}{d\theta_1},$$

or $\qquad -h \sin i \frac{di}{dt} + \cos i \frac{dh}{dt} = \frac{dR}{d\theta_1};$

therefore $\quad h \sin i \frac{di}{dt} = \frac{dR}{d\theta_1} - \cos i \frac{dh}{dt}$

$$= \frac{dR}{d\theta_1} - \cos i \frac{dR}{d\theta}$$

$$= \frac{dR}{d\Omega} + (1 - \cos i)\frac{dR}{d\theta}, \text{ (Art. 16),}$$

$$= \frac{dR}{d\Omega} + (1 - \cos i)\left(\frac{dR}{d\epsilon} + \frac{dR}{d\varpi}\right), \text{ (Art. 15);}$$

therefore $\dfrac{di}{dt} = -\dfrac{1}{h \sin i}\left\{\dfrac{dR}{d\Omega} + (1 - \cos i)\left(\dfrac{dR}{d\epsilon} + \dfrac{dR}{d\varpi}\right)\right\}$

$$= -\frac{na}{\mu \sqrt{(1 - e^2)}}\left\{\frac{1}{\sin i}\frac{dR}{d\Omega} + \tan\frac{i}{2}\left(\frac{dR}{d\epsilon} + \frac{dR}{d\varpi}\right)\right\}.$$

35. *To obtain a formula for calculating the longitude of the node.* (*Second method.*)

We have by Art. 13,

$$R = f(r, \theta, \Omega, i) \dots\dots\dots\dots(1),$$

or, since evidently $\qquad \theta - \Omega = \theta_0 - \Omega_0,$

$$R = f(r, \theta_0 + \Omega - \Omega_0, \Omega, i)* \dots\dots \dots(2),$$

---

* We have made this transformation, because, although the value of $\dfrac{d\theta_0}{dt}$ is the same in form as if the elements were invariable, this is not the case with $\dfrac{d\theta}{dt}$.

Now we have seen (Art. 26) that $\dfrac{d\,(R)}{dt}$ (denoting by it that R is to be differentiated with respect to t only so far as it involves t through involving the position of the disturbed planet) may be expressed in the same form as if the elements were invariable.

We have then, considering the elements variable,

$$\frac{d\,(R)}{dt} = \frac{dR}{dr}\frac{dr}{dt} + \frac{dR}{d\theta_0}\frac{d\,(\theta_0+\Omega-\Omega_0)}{dt} + \frac{dR}{d\Omega}\frac{d\Omega}{dt} + \frac{dR}{di}\frac{di}{dt},$$

and, considering them invariable,

$$\frac{d\,(R)}{dt} = \frac{dR}{dr}\frac{dr}{dt} + \frac{dR}{d\theta_0}\frac{d\theta_0}{dt}.$$

Equating the two values of $\dfrac{d\,(R)}{dt}$, we obtain

$$\frac{dR}{d\theta_0}\left(\frac{d\Omega}{dt} - \frac{d\Omega_0}{dt}\right) + \frac{dR}{d\Omega}\frac{d\Omega}{dt} + \frac{dR}{di}\frac{di}{dt} = 0.$$

Now $\dfrac{dR}{d\theta_0}$ in (2) $= \dfrac{dR}{d\theta}$ in (1) $= \dfrac{dR}{d\epsilon} + \dfrac{dR}{d\varpi}$,

and (see Art. 29), $\dfrac{d\Omega_0}{dt} = \dfrac{d\Omega}{dt}\cos i$;

therefore $\left\{\dfrac{dR}{d\Omega} + (1 - \cos i)\left(\dfrac{dR}{d\epsilon} + \dfrac{dR}{d\varpi}\right)\right\}\dfrac{d\Omega}{dt} + \dfrac{dR}{di}\dfrac{di}{dt} = 0.$

Substituting for $\dfrac{di}{dt}$ its value,

$$\frac{d\Omega}{dt} = \frac{na}{\mu\sqrt{(1 - e^2)}\sin i}\frac{dR}{di}.$$

36. *To obtain a formula for calculating the longitude of the epoch.*

If R be expressed in terms of t and the elements (see Art. 13), since $nt + \epsilon$ always occurs as one symbol, we may write

$$R = f\,(nt + \epsilon,\, a,\, e,\, \varpi,\, \Omega,\, i).$$

Differentiating, the elements being considered variable, we have

$$\frac{d(R)}{dt} = \frac{dR}{d\epsilon}\frac{d(nt+\epsilon)}{dt} + \frac{dR}{da}\frac{da}{dt} + \frac{dR}{de}\frac{de}{dt} + \frac{dR}{d\varpi}\frac{d\varpi}{dt}$$
$$+ \frac{dR}{d\Omega}\frac{d\Omega}{dt} + \frac{dR}{di}\frac{di}{dt},$$

and differentiating as if the elements were invariable, which is permissible for the reason explained in Art. 26,

$$\frac{d(R)}{dt} = n\frac{dR}{d\epsilon}.$$

Equating the two values of $\frac{d(R)}{dt}$,

$$n\frac{dR}{d\epsilon} = \frac{dR}{d\epsilon}\left(n + t\frac{dn}{dt} + \frac{d\epsilon}{dt}\right) + \frac{dR}{da}\frac{da}{dt} + \frac{dR}{de}\frac{de}{dt} + \frac{dR}{d\varpi}\frac{d\varpi}{dt}$$
$$+ \frac{dR}{d\Omega}\frac{d\Omega}{dt} + \frac{dR}{di}\frac{di}{dt}.$$

Substituting for $\frac{da}{dt}$, $\frac{de}{dt}$, &c., their values

$$0 = \frac{dR}{d\epsilon}\left(t\frac{dn}{dt} + \frac{d\epsilon}{dt}\right) + \frac{2na^2}{\mu}\frac{dR}{d\epsilon}\frac{dR}{da} + \frac{na(1-e^2)}{\mu e}\frac{dR}{d\epsilon}\frac{dR}{de}$$

$$- \frac{na\sqrt{(1-e^2)}}{\mu e}\left(\frac{dR}{d\epsilon} + \frac{dR}{d\varpi}\right)\frac{dR}{de} + \frac{na\sqrt{(1-e^2)}}{\mu e}\frac{dR}{de}\frac{dR}{d\varpi}$$

$$+ \frac{na\tan\frac{i}{2}}{\mu\sqrt{(1-e^2)}}\frac{dR}{di}\frac{dR}{d\varpi} + \frac{na}{\mu\sqrt{(1-e^2)}\sin i}\frac{dR}{di}\frac{dR}{d\Omega}$$

$$- \frac{na}{\mu\sqrt{(1-e^2)}\sin i}\frac{dR}{d\Omega}\frac{dR}{di} - \frac{na\tan\frac{i}{2}}{\mu\sqrt{(1-e^2)}}\left(\frac{dR}{d\epsilon} + \frac{dR}{d\varpi}\right)\frac{dR}{di},$$

C. P. T.

3

$$= \frac{dR}{d\epsilon}\left(t\frac{dn}{dt} + \frac{d\epsilon}{dt}\right) + \frac{2na^2}{\mu}\frac{dR}{d\epsilon}\frac{dR}{da}$$

$$- \frac{na\sqrt{(1-e^2)}}{\mu e}\{1 - \sqrt{(1-e^2)}\}\frac{dR}{d\epsilon}\frac{dR}{de}$$

$$- \frac{na\tan\frac{i}{2}}{\mu\sqrt{(1-e^2)}}\frac{dR}{d\epsilon}\frac{dR}{di}.$$

Dividing every term by $\dfrac{dR}{d\epsilon}$, and transposing, we obtain

$$\frac{d\epsilon}{dt} = -t\frac{dn}{dt} - \frac{2na^2}{\mu}\frac{dR}{da} + \frac{na\sqrt{(1-e^2)}}{\mu e}\{1 - \sqrt{(1-e^2)}\}\frac{dR}{de}$$

$$+ \frac{na\tan\frac{i}{2}}{\mu\sqrt{(1-e^2)}}\frac{dR}{di}.$$

37. Of the formulæ which have been obtained for calculating the elements of the orbit, that of the preceding Article is the only one which contains a term proportional to the time*. It may, however, be replaced by one in which no such term exists. For, let ξ denote the mean longitude, then

$$\xi = nt + \epsilon;$$

therefore

$$\frac{d\xi}{dt} = n + t\frac{dn}{dt} + \frac{d\epsilon}{dt},$$

or

$$\frac{d}{dt}(\xi - \int n dt) = t\frac{dn}{dt} + \frac{d\epsilon}{dt}.$$

Now let

$$\xi = \int n dt + \epsilon',$$

then by the formula of the last Article

$$\frac{d\epsilon'}{dt} = -\frac{2na^2}{\mu}\frac{dR}{da} + \frac{na\sqrt{(1-e^2)}}{\mu e}\{1 - \sqrt{(1-e^2)}\}\frac{dR}{de}$$

$$+ \frac{na\tan\frac{i}{2}}{\mu\sqrt{(1-e^2)}}\frac{dR}{di}.$$

---

* The reader, if acquainted with the Lunar Theory, will have already seen the inconvenience of such terms.

Since in the elliptic formulæ ϵ never occurs except when coupled with nt, in the expression $nt + \epsilon$, it will be altogether eliminated if for $nt + \epsilon$ we write $\int n dt + \epsilon'$.

Considered as replacing the element ϵ, ϵ' is called the *epoch**. Since however we shall never have occasion to employ the formula of the preceding Article, the accent will in future be omitted.

$\int n dt$ is termed the *mean motion* in the disturbed orbit, and is denoted by ζ.

38. To obtain a formula for calculating ζ, we have

$$\frac{d^2\zeta}{dt^2} = \frac{dn}{dt},$$

and differentiating the equation $n^2 a^3 = \mu$,

$$2na^3 \frac{dn}{dt} + 3n^2 a^2 \frac{da}{dt} = 0;$$

therefore

$$\frac{dn}{dt} = -\frac{3n}{2a}\frac{da}{dt} = -\frac{3n^2 a}{\mu}\frac{dR}{d\epsilon},$$

or

$$\frac{d^2\zeta}{dt^2} = -\frac{3n^2 a}{\mu}\frac{dR}{d\epsilon}.$$

39. We will here recapitulate the formulæ which have been obtained for calculating the elements of the orbit:

(i) $\dfrac{da}{dt} = \dfrac{2na^2}{\mu}\dfrac{dR}{d\epsilon},$

(ii) $\dfrac{de}{dt} = \dfrac{na(1-e^2)}{\mu e}\dfrac{dR}{d\epsilon} - \dfrac{na\sqrt{(1-e^2)}}{\mu e}\left(\dfrac{dR}{d\epsilon} + \dfrac{dR}{d\varpi}\right),$

(iii) $\dfrac{d\varpi}{dt} = \dfrac{na\sqrt{(1-e^2)}}{\mu e}\dfrac{dR}{d\epsilon} + \dfrac{na\tan\frac{i}{2}}{\mu\sqrt{(1-e^2)}}\dfrac{dR}{di},$

---

* It may be noticed that if n were constant, ϵ' would be identical with ϵ.

3—2

(iv) $\dfrac{d\epsilon}{dt} = -\dfrac{2na^2}{\mu}\dfrac{dR}{da} + \dfrac{na\sqrt{(1-e^2)}}{\mu e}\{1 - \sqrt{(1-e^2)}\}\dfrac{dR}{de}$

$$+ \dfrac{na\tan\dfrac{i}{2}}{\mu\sqrt{(1-e^2)}}\dfrac{dR}{di},$$

(v) $\dfrac{d\Omega}{dt} = \dfrac{na}{\mu\sqrt{(1-e^2)}\sin i}\dfrac{dR}{di},$

(vi) $\dfrac{di}{dt} = -\dfrac{na}{\mu\sqrt{(1-e^2)}}\left\{\dfrac{1}{\sin i}\dfrac{dR}{d\Omega} + \tan\dfrac{i}{2}\left(\dfrac{dR}{d\epsilon} + \dfrac{dR}{d\varpi}\right)\right\}.$

We have also (vii) the equation

$$\dfrac{d^2\zeta}{dt^2} = -\dfrac{3n^2a}{\mu}\dfrac{dR}{d\epsilon},$$

but this forms no new relation, since it has been deduced from (i).

40.   When the elements have been calculated by means of the above formulæ, the position of the planet will be given by the equations

$$r = a\left\{1 + \dfrac{1}{2}e^2 - e\cos(\zeta + \epsilon - \varpi) - \dfrac{1}{2}e^2\cos 2(\zeta + \epsilon - \varpi) - \ldots\right\},$$

$$\theta = \zeta + \epsilon + 2e\sin(\zeta + \epsilon - \varpi) + \dfrac{5}{4}e^2\sin 2(\zeta + \epsilon - \varpi) + \ldots.$$

Note. Some of the Author's methods are similar to those given by Mr M. O'Brien in an article "*On certain formulæ in Physical Astronomy*," published in the *Camb. Math. Journal*, 1843, Vol. III. pp. 249—259. Compare sections 18, 29, 30, 8, 20, and 31 of this work with sections 6, 7, 8, 10, 11 of the article referred to. [A. F.]

CHAPTER III.

DEVELOPMENT OF THE DISTURBING FUNCTION.

41. In the first Chapter we have obtained equations by means of which R may be expressed in terms of the time and the elements of the orbit; we now proceed to shew how the actual development may be effected in a series ascending by powers and products of the excentricities and the tangents of the inclinations. In the Planetary Theory these are extremely small, and the series will converge rapidly. Accordingly in the present treatise small quantities of orders higher than the second will be neglected*.

42. If we recur to Art. 11, it will be seen that, considering only one disturbing planet,

$$R = m'\left[\frac{1}{\{r_1^2 + r_1'^2 - 2r_1 r_1' \cos(\theta_1 - \theta_1') + (z - z')^2\}^{\frac{1}{2}}} - \frac{r_1 r_1' \cos(\theta_1 - \theta_1') + zz'}{(r_1'^2 + z'^2)^{\frac{3}{2}}}\right].$$

The first step towards the required development will be the expansion of r_1, r_1', θ_1, θ_1', z and z' in terms of the time

* We may remark that to this order of approximation the inclinations, their sines, and tangents will be equal.

and the elements of the orbit. For this purpose we may employ the equations which have already been obtained in Art. 13, viz.:

$$\tan (\theta_1 - \Omega) = \cos i \tan (\theta - \Omega),$$

$$\sin \lambda = \sin i \sin (\theta - \Omega),$$

$$r_1 = r \cos \lambda, \qquad z = r \sin \lambda,$$

$$r = a \left\{ 1 + \frac{1}{2} e^2 - e \cos (nt + \epsilon - \varpi) - \frac{1}{2} e^2 \cos 2 (nt + \epsilon - \varpi) - \ldots \right\},$$

$$\theta = nt + \epsilon + 2e \sin (nt + \epsilon - \varpi) + \frac{5}{4} e^2 \sin 2 (nt + \epsilon - \varpi) + \ldots,$$

with similar equations involving the co-ordinates and elements of the disturbing planet.

(i) To expand r_1. We have

$$r_1 = r \cos \lambda = r (1 - \sin^2 \lambda)^{\frac{1}{2}}$$

$$= r \left(1 - \frac{1}{2} \sin^2 \lambda + \ldots \right)$$

$$= r \left\{ 1 - \frac{1}{2} \sin^2 i \sin^2 (\theta - \Omega) + \ldots \right\}$$

$$= r \left\{ 1 - \frac{1}{2} \tan^2 i \sin^2 (\theta - \Omega) + \ldots \right\}$$

to the same order of approximation,

$$= r \left\{ 1 - \frac{1}{4} \tan^2 i + \frac{1}{4} \tan^2 i \cos 2 (\theta - \Omega) + \ldots \right\};$$

or substituting the expansions for r and θ,

$$r_1 = a \left\{ 1 + \frac{1}{2} e^2 - \frac{1}{4} \tan^2 i - e \cos (nt + \epsilon - \varpi) - \frac{1}{2} e^2 \cos 2 (nt + \epsilon - \varpi) \right.$$
$$\left. + \frac{1}{4} \tan^2 i \cos 2 (nt + \epsilon - \Omega) + \ldots \right\}$$

$$= a (1 + u), \text{ suppose.}$$

Similarly, $\qquad\qquad r_1' = a' (1 + u').$

(ii)   To expand θ_1.   We have

$$\tan(\theta_1 - \theta) = \tan\{(\theta_1 - \Omega) - (\theta - \Omega)\}$$

$$= \frac{\tan(\theta_1 - \Omega) - \tan(\theta - \Omega)}{1 + \tan(\theta_1 - \Omega)\tan(\theta - \Omega)}$$

$$= \frac{(\cos i - 1)\tan(\theta - \Omega)}{1 + \cos i \tan^2(\theta - \Omega)}$$

$$= \frac{-2\sin^2\frac{i}{2}\tan(\theta - \Omega)}{1 + \tan^2(\theta - \Omega) - 2\sin^2\frac{i}{2}\tan^2(\theta - \Omega)}$$

$$= \frac{-\sin^2\frac{i}{2}\sin 2(\theta - \Omega)}{1 - 2\sin^2\frac{i}{2}\sin^2(\theta - \Omega)}$$

$$= -\sin^2\frac{i}{2}\sin 2(\theta - \Omega) - \ldots\,;$$

therefore $\theta_1 - \theta = -\sin^2\frac{i}{2}\sin 2(\theta - \Omega) - \ldots$

$$= -\tan^2\frac{i}{2}\sin 2(\theta - \Omega) - \ldots\,,$$

to the same order of approximation; or, substituting the expansion for θ,

$$\theta_1 = nt + \epsilon + 2e\sin(nt + \epsilon - \varpi) + \frac{5}{4}e^2\sin 2(nt + \epsilon - \varpi)$$

$$- \tan^2\frac{i}{2}\sin 2(nt + \epsilon - \Omega) + \ldots$$

$= nt + \epsilon + v$, suppose.

Similarly, $\theta_1' = n't + \epsilon' + v'$.

(iii) To expand z. We have

$$z = r \sin \lambda = r \sin i \sin (\theta - \Omega)$$

$$= r \tan i \sin (\theta - \Omega) - \dots$$

to the second order ; or, substituting the expansions for r and θ,

$$z = a \{\tan i \sin (nt + \epsilon - \Omega) + \dots\},$$

in which terms of the second order will not be required.

A similar expression may be found for z'.

43. Having obtained the expansions of r_1, r_1', θ_1, θ_1', z, z' we must now substitute them in the expression for R. This may be effected as follows.

Let $R_,$ be the value of R when u, u', v, v' are severally zero : then, writing ϕ for $nt + \epsilon - (n't + \epsilon')$ we have

$$R_, = m' \left[\{a^2 + a'^2 - 2aa' \cos \phi + (z - z')^2\}^{-\frac{1}{2}} \right.$$

$$\left. - (aa' \cos \phi + zz') (a'^2 + z'^2)^{-\frac{3}{2}} \right]$$

$$= m' \left[(a^2 + a'^2 - 2aa' \cos \phi)^{-\frac{1}{2}} - \frac{a}{a'^2} \cos \phi \right]$$

$$- m' \left[\frac{1}{2} (a^2 + a'^2 - 2aa' \cos \phi)^{-\frac{3}{2}} (z - z')^2 \right.$$

$$\left. + \frac{1}{a'^3} zz' - \frac{3}{2} \frac{a}{a'^4} z'^2 \cos \phi \right]$$

$$+ \dots\dots\dots\dots$$

Now $R = R_, + \dfrac{dR_,}{da} au + \dfrac{dR_,}{da'} a'u' + \dfrac{dR_,}{d\phi} (v - v')$

$$+ \frac{1}{2} \left\{ \frac{d^2 R_,}{da^2} a^2 u^2 + \frac{d^2 R_,}{da'^2} a'^2 u'^2 + \frac{d^2 R_,}{d\phi^2} (v - v')^2 \right\}$$

$$+ \frac{d^2 R_,}{da \, da'} aa' uu' + \frac{d^2 R_,}{da \, d\phi} au (v - v') + \frac{d^2 R_,}{da' \, d\phi} a'u' (v - v')$$

$$+ \dots\dots\dots\dots .$$

44. It will be shewn in a subsequent Article that

$$(a^2 + a'^2 - 2aa' \cos \phi)^{-s}$$

can be expanded in a series of the form

$$\frac{1}{2} A_0 + A_1 \cos \phi + A_2 \cos 2\phi + \ldots + A_k \cos k\phi + \ldots$$

Assume then

$$(a^2 + a'^2 - 2aa' \cos \phi)^{-\frac{1}{2}} = \frac{1}{2} C_0 + C_1 \cos \phi + C_2 \cos 2\phi + \ldots$$

$$(a^2 + a'^2 - 2aa' \cos \phi)^{-\frac{s}{2}} = \frac{1}{2} D_0 + D_1 \cos \phi + D_2 \cos 2\phi + \ldots$$

Thus

$$R = m' \left\{ \frac{1}{2} C_0 + \left(C_1 - \frac{a}{a'^2} \right) \cos \phi + C_2 \cos 2\phi + \ldots \right\}$$

$$+ m'au \left\{ \frac{1}{2} \frac{dC_0}{da} + \left(\frac{dC_1}{da} - \frac{1}{a'^2} \right) \cos \phi + \frac{dC_2}{da} \cos 2\phi + \ldots \right\}$$

$$+ m'a'u' \left\{ \frac{1}{2} \frac{dC_0}{da'} + \left(\frac{dC_1}{da'} + \frac{2a}{a'^3} \right) \cos \phi + \frac{dC_2}{da'} \cos 2\phi + \ldots \right\}$$

$$- m' (v - v') \left\{ \left(C_1 - \frac{a}{a'^2} \right) \sin \phi + 2C_2 \sin 2\phi + \ldots \right\}$$

$$+ \frac{m'a^2u^2}{2} \left\{ \frac{1}{2} \frac{d^2C_0}{da^2} + \frac{d^2C_1}{da^2} \cos \phi + \frac{d^2C_2}{da^2} \cos 2\phi + \ldots \right\}$$

$$+ \frac{m'a'^2u'^2}{2} \left\{ \frac{1}{2} \frac{d^2C_0}{da'^2} + \left(\frac{d^2C_1}{da'^2} - \frac{6a}{a'^4} \right) \cos \phi + \frac{d^2C_2}{da'^2} \cos 2\phi + \ldots \right\}$$

$$- \frac{m' (v - v')^2}{2} \left\{ \left(C_1 - \frac{a}{a'^2} \right) \cos \phi + 4C_2 \cos 2\phi + \ldots \right\}$$

$$+ m'aa'uu' \left\{ \frac{1}{2} \frac{d^2C_0}{da\,da'} + \left(\frac{d^2C_1}{da\,da'} + \frac{2}{a'^3} \right) \cos \phi + \frac{d^2C_2}{da\,da'} \cos 2\phi + \ldots \right\}$$

$$- m'au (v - v') \left\{ \left(\frac{dC_1}{da} - \frac{1}{a'^2} \right) \sin \phi + 2 \frac{dC_2}{da} \sin 2\phi + \ldots \right\}$$

$$- m'a'u'(v - v') \left\{ \left(\frac{dC_1}{da'} + \frac{2a}{a'^3} \right) \sin \phi + 2 \frac{dC_2}{da'} \sin 2\phi + ... \right\}$$

$$- \frac{m'(z - z')^2}{2} \left\{ \frac{1}{2} D_0 + D_1 \cos \phi + D_2 \cos 2\phi + ... \right\}$$

$$- m' \left\{ \frac{zz'}{a'^3} - \frac{3}{2} \frac{a}{a'^4} z'^2 \cos \phi + ... \right\}.$$

45. By Art. 42,

$$u = \frac{1}{2} e^2 - \frac{1}{4} \tan^2 i - e \cos(nt + \epsilon - \varpi) - \frac{1}{2} e^2 \cos 2(nt + \epsilon - \varpi)$$

$$+ \frac{1}{4} \tan^2 i \cos 2(nt + \epsilon - \Omega) + ...$$

$$v = 2e \sin(nt + \epsilon - \varpi) + \frac{5}{4} e^2 \sin 2(nt + \epsilon - \varpi)$$

$$- \tan^2 \frac{i}{2} \sin 2(nt + \epsilon - \Omega) + ...,$$

$$z = a \{ \tan i \sin(nt + \epsilon - \Omega) + ... \},$$

with similar expressions for u', v', z'.

Hence $u^2 = e^2 \cos^2(nt + \epsilon - \varpi) + ...$

$$= \frac{e^2}{2} + \frac{e^2}{2} \cos 2(nt + \epsilon - \varpi) + ...,$$

$$(v - v')^2 = 4e^2 \sin^2(nt + \epsilon - \varpi) + 4e'^2 \sin^2(n't + \epsilon' - \varpi')$$

$$- 8ee' \sin(nt + \epsilon - \varpi) \sin(n't + \epsilon' - \varpi') + ...$$

$$= 2(e^2 + e'^2) - 2e^2 \cos 2(nt + \epsilon - \varpi) - 2e'^2 \cos 2(n't + \epsilon' - \varpi')$$

$$- 4ee' \cos(\phi - \varpi + \varpi') + 4ee' \cos\{(n + n')t + \epsilon + \epsilon' - \varpi - \varpi'\} + ...,$$

$$uu' = ee' \cos(nt + \epsilon - \varpi) \cos(n't + \epsilon' - \varpi') + ...$$

$$= \frac{ee'}{2} \cos(\phi - \varpi + \varpi') + \frac{ee'}{2} \cos\{(n + n')t + \epsilon + \epsilon' - \varpi - \varpi'\} + ...$$

$$u\,(v-v') = -\,e^{2}\sin 2\,(nt+\epsilon-\varpi)$$
$$+\,2ee'\cos\,(nt+\epsilon-\varpi)\sin\,(n't+\epsilon'-\varpi')+\dots$$
$$= -\,e^{2}\sin 2\,(nt+\epsilon-\varpi) - ee'\sin\,(\phi-\varpi+\varpi')$$
$$+\,ee'\sin\,\{(n+n')\,t+\epsilon+\epsilon'-\varpi-\varpi'\}+\dots,$$

$$(z-z')^{2} = \frac{a^{2}\tan^{2}i}{2} + \frac{a'^{2}\tan^{2}i'}{2} - \frac{a^{2}\tan^{2}i}{2}\cos 2\,(nt+\epsilon-\Omega)$$

$$-\,\frac{a'^{2}\tan^{2}i'}{2}\cos 2\,(n't+\epsilon'-\Omega') - aa'\tan i\tan i'\cos(\phi-\Omega+\Omega')$$

$$+\,aa'\tan i\tan i'\cos\,\{(n+n')\,t+\epsilon+\epsilon'-\Omega-\Omega'\}+\dots$$

&c.

46.   If these values be substituted in Art. 44, it will be seen that cosines will bu multiplied only by cosines, and sines by sines.   Hence the series will consist of two parts, one inde-pendent of t explicitly, and the other consisting of periodical terms of the form

$$P\cos\,\{(pn\pm qn')\,t+Q\},$$

where p and q are any positive integers or zero, P is a function of the mean distances, excentricities, and inclinations, and Q a function of the longitudes of perihelia, nodes, and epochs. The former part is denoted by the symbol F: we proceed to determine its value as far as the second order of small quan-tities.

47.   *To determine that part of* R *which is independent of the time explicitly.*

If those terms only be written down which either are, or after reduction will become, independent of t, we have

$$m'\left\{\frac{C_{0}}{2} + \frac{a}{2}\frac{dC_{0}}{da}\left(\frac{e^{2}}{2} - \frac{\tan^{2}i}{4}\right) + \frac{a'}{2}\frac{dC_{0}}{da'}\left(\frac{e'^{2}}{2} - \frac{\tan^{2}i'}{4}\right)\right.$$

$$+\,\frac{a^{2}}{4}\frac{d^{2}C_{0}}{da^{2}}\frac{e^{2}}{2} + \frac{a'^{2}}{4}\frac{d^{2}C_{0}}{da'^{2}}\frac{e'^{2}}{2} - \frac{D_{0}}{4}\left(\frac{a^{2}\tan^{2}i}{2} + \frac{a'^{2}\tan^{2}i'}{2}\right)$$

$$+ \frac{1}{2}\left(C_1 - \frac{a}{a'^2}\right) \cos \phi \; 4ee' \cos(\phi - \varpi + \varpi')$$

$$+ aa'\left(\frac{d^2C_1}{da\,da'} + \frac{2}{a'^3}\right) \cos \phi \; \frac{ee'}{2} \cos(\phi - \varpi + \varpi')$$

$$+ a\left(\frac{dC_1}{da} - \frac{1}{a'^2}\right) \sin \phi \; ee' \sin(\phi - \varpi + \varpi')$$

$$+ a'\left(\frac{dC_1}{da'} + \frac{2a}{a'^3}\right) \sin \phi \; ee' \sin(\phi - \varpi + \varpi')$$

$$+ \frac{D_1}{2} \cos \phi \; aa' \tan i \tan i'' \cos(\phi - \Omega + \Omega') + \dots \bigg\}.$$

Now $\cos \phi \cos(\phi - \varpi + \varpi')$ and $\sin \phi \sin(\phi - \varpi + \varpi')$ contain the term $\frac{1}{2}\cos(\varpi - \varpi')$, $\cos \phi \cos(\phi - \Omega + \Omega')$ contains the term $\frac{1}{2}\cos(\Omega - \Omega')$; hence

$$F = m'\left\{\frac{C_0}{2} + \frac{1}{4}\left(a\,\frac{dC_0}{da} + \frac{a^2}{2}\frac{d^2C_0}{da^2}\right)e^2\right.$$

$$+ \frac{1}{4}\left(a'\frac{dC_0}{da'} + \frac{a'^2}{2}\frac{d^2C_0}{da'^2}\right)e'^2$$

$$+ \frac{1}{4}\left(4C_1 + 2a\,\frac{dC_1}{da} + 2a'\frac{dC_1}{da'} + aa'\frac{d^2C_1}{da\,da'}\right)ee'\cos(\varpi - \varpi')$$

$$- \frac{1}{8}\left(a^2D_0 + a\,\frac{dC_0}{da}\right)\tan^2 i - \frac{1}{8}\left(a'^2D_0 + a'\frac{dC_0}{da'}\right)\tan^2 i''$$

$$+ \frac{1}{4}aa'D_1 \tan i \tan i'' \cos(\Omega - \Omega') + \dots \bigg\}.$$

We shall hereafter be able to simplify this expression.

48. We have seen that the remaining terms of R are of the form $P \cos\{(pn \pm qn')t + Q\}$: if then values of p and q could be found such that $pn \pm qn' = 0$, this term, being in-

dependent of t explicitly, would form an additional term in F. No instance of this, however, occurs among the planets.

49.   In consequence of the extreme smallness of the ex-centricities and inclinations of the orbits of the principal planets, terms in R of orders higher than the second may in general be neglected : but it sometimes happens, as in the Lunar Theory, that higher terms become sensible through the process of integration.   This we shall consider in a subsequent Chapter, but the following proposition has an important bearing on the subject.

50.   *The principal part of the coefficient of a term in* R *of the form* P cos {(pn − qn′) t + Q} *is of the order* p ∼ q.

DEF.   By the principal part of the coefficient is meant that part of P which is of lowest dimensions in e, e', $\tan i$, $\tan i'$.

If we return to the expression for R in Art. 44, it will be seen that in order to obtain the general term it will be necessary to multiply the product of the general terms of the expansions for u^a, u'^β, v^γ, v'^δ, z^ϵ, z'^ζ by $\cos k\phi$ or $\sin k\phi$.

Now (i) in the expansions of u, u', v, v', z, z' the following law is observed to hold :—The number which multiplies $nt + \epsilon$ or $n't + \epsilon'$ in the argument of any term represents the order of the principal part of the coefficient of that term.

(2)   The same holds good in any power of u, u', v, v', z, or z'.   For consider a term $P \cos (pnt + q)$ in u^2.   It can only have arisen in the following ways; partly from the multiplication of two terms in u of which the arguments are $lnt + \lambda$ and $mnt + \mu$, where $l + m = p$; and partly from such as have the arguments $l'nt + \lambda'$ and $m'nt + \mu'$, where $l' \sim m' = p$. In the former case the order of the coefficient will be $l + m$, which equals p, in the latter it will be $l' + m'$, and this is

greater than p. Hence the principal part of the coefficient of a term $P \cos (pnt + q)$ in u^2, will be of the order p.

Since then the law holds in u^2, it may be shewn in like manner to hold in the product of u^2 and u, i.e. in u^3. Thus it may be proved for any power of u. In like manner it may be shewn to hold for any powers of u', v, v', z or z'.

(3) The same law is true for the product of any powers of u, v, z; and likewise for the product of any powers of u', v', z'. This may be proved by a method similar to that of (2).

(4) In the product of any powers of u, u', v, v', z and z', the order of the principal part of the coefficient is the arithmetical sum of the multipliers of nt and $n't$.

For let us consider a term $M \cos \{(ln \pm l'n') t + N\}$. Now this must evidently have arisen from the multiplication of $L \cos (lnt + \lambda)$ with $L' \cos (l'n't + \lambda')$, or of $L \sin (lnt + \lambda)$ with $L' \sin (l'n't + \lambda')$, where by (3) L is of the order l and L' of the order l'. Hence M will be of the order $l + l'$.

Now any term in the development of R of the form $P \cos \{(pn - qn') t + Q\}$ must have arisen partly from the multiplication of $P_1 {\cos \atop \sin} k\phi$, or as it may be written

$$P_1 {\cos \atop \sin} \{(kn - kn') t + Q_1\}$$

with
$$P_2 {\cos \atop \sin} \{[(p - k) n - (q - k) n'] t + Q_2\},$$

and partly from its multiplication with

$$P_3 {\cos \atop \sin} \{[(p + k) n - (q + k) n'] t + Q_3\},$$

where k is any positive integer or zero, P_1 is a function of a and a' only, and P_2, P_3 are functions of the excentricities and inclinations, such that the orders of their principal parts are given by law (4). Hence the order of the principal part of P will be equal to the lesser of those of P_2 and P_3.

Now the order of the principal part of P_2 will be the least value of which the arithmetical sum of $p \sim k$ and $q \sim k$ is susceptible, for different values of k.

(i)   Suppose k intermediate to p and q; then this sum

$$= p \sim k + q \sim k = p \sim q;$$

(ii)   Suppose k not greater than the smaller of p and q; then this sum $= p + q - 2k$, the least value of which (by putting k equal to the smaller of p and q) $= p \sim q$;

(iii)   Suppose k not less than the greater of p and q; then this sum $= 2k - p - q$, the least value of which (by putting k equal to the greater of p and q) $= p \sim q$.

Thus $p \sim q$ is the order of the principal part of P_2. That of P_3 will be the least value of which $p + k + q + k$ is susceptible, i.e. $p + q$.

Hence it appears that the order of the principal part of P is $p \sim q$.

51.   *The principal part of the coefficient of a term in R, of the form* P cos $\{(pn + qn')t + Q\}$ *is of the order* p + q.

This term arises from the multiplication of such terms as

$$P_1 \begin{smallmatrix} \cos \\ \sin \end{smallmatrix} \{(kn - kn')t + Q_1\},$$

with $\qquad P_2 \begin{smallmatrix} \cos \\ \sin \end{smallmatrix} \{[(p - k)n + (q + k)n']t + Q_2\},$

and $\qquad P_3 \begin{smallmatrix} \cos \\ \sin \end{smallmatrix} \{[(p + k)n + (q - k)n']t + Q_3\},$

and as in the last Article, the order of the principal part of P will be equal to the lesser of those of P_2 and P_3.

Now the order of the principal part of P_2 will be the least value which the arithmetical sum of $p \sim k$ and $q + k$ can assume, for different values of k.

(i) Suppose k less than p; then this sum

$$= p - k + q + k = p + q.$$

(ii) Suppose k not less than p; then this sum

$$= k - p + q + k,$$

the least value of which (by putting k equal to p) $= p + q$.

Similarly it may be shewn that $p + q$ will be the order of the principal part of P_3.

Hence it follows that $p + q$ will be the order of the principal part of P_3.

In Art. 44 we have assumed that $(a^2 + a'^2 - 2aa' \cos \phi)^{-s}$ can be expanded in a series of cosines of ϕ and its multiples, we shall now give a proof of this and shew how the coefficients may be calculated.

52.　*To shew that* $(a^2 + a'^2 - 2aa' \cos \phi)^{-s}$ *can be expanded in a series of cosines of multiples of* ϕ.

Suppose a greater than a', and for $\dfrac{a'}{a}$ write α; then

$$(a^2 + a'^2 - 2aa' \cos \phi)^{-s} = a^{-2s} (1 + \alpha^2 - 2\alpha \cos \phi)^{-s}$$

$$= a^{-2s} \{1 + \alpha^2 - \alpha (e^{\phi \sqrt{-1}} + e^{-\phi \sqrt{-1}})\}^{-s}$$

$$= a^{-2s} (1 - \alpha e^{\phi \sqrt{-1}})^{-s} (1 - \alpha e^{-\phi \sqrt{-1}})^{-s}$$

$$= a^{-2s} \left\{ 1 + s\alpha e^{\phi\sqrt{-1}} + \frac{s(s+1)}{\underline{2}} \alpha^2 e^{2\phi\sqrt{-1}} \right.$$

$$\left. + \frac{s(s+1)(s+2)}{\underline{3}} \alpha^3 e^{3\phi\sqrt{-1}} + \ldots \right\}$$

$$\times \left\{ 1 + s\alpha e^{-\phi\sqrt{-1}} + \frac{s(s+1)}{2} \alpha^2 e^{-2\phi\sqrt{-1}} \right.$$

$$\left. + \frac{s(s+1)(s+2)}{\underline{3}} \alpha^2 e^{-3\phi\sqrt{-1}} + \ldots \right\}$$

$$= a^{-2s} \left[1 + s^2\alpha^2 + \left\{ \frac{s(s+1)}{\underline{2}} \right\}^2 \alpha^4 \right.$$

$$\left. + \left\{ \frac{s(s+1)(s+2)}{\underline{3}} \right\}^2 \alpha^6 + \ldots \right]$$

$$+ 2a^{-2s} \left\{ s\alpha + s . \frac{s(s+1)}{\underline{2}} \alpha^3 \right.$$

$$\left. + \frac{s(s+1)}{\underline{2}} . \frac{s(s+1)(s+2)}{\underline{3}} \alpha^5 + \ldots \right\} \left(\frac{e^{\phi\sqrt{-1}} + e^{\phi\sqrt{-1}}}{2} \right)$$

$$+ \ldots \ldots \ldots \ldots$$

where the coefficients of $e^{k\phi\sqrt{-1}}$ and $e^{-k\phi\sqrt{-1}}$ will always be equal. Hence we may write

$$(a^2 + a'^2 - 2aa' \cos \phi)^{-s}$$

$$= \frac{1}{2} A_0 + A_1 \cos \phi + A_2 \cos 2\phi + \ldots + A_k \cos k\phi + \ldots,$$

where A_0, A_1, &c., are functions of a and a'. The series which they represent will be always convergent provided α is less than unity, or a greater than a'. If a be less than a', we have only to interchange a and a' in the above, so that α will then denote the ratio of a to a'.

53.   *To calculate* C_0 *and* C_1.

In the preceding Article, let $s = \dfrac{1}{2}$; then

$$C_0 = 2a^{-1}\left\{1 + \left(\frac{1}{2}\right)^2 a^2 + \left(\frac{1 \cdot 3}{2 \cdot 4}\right)^2 a^4 + \left(\frac{1 \cdot 3 \cdot 5}{2 \cdot 4 \cdot 6}\right)^2 a^6 + \ldots\right\},$$

$$C_1 = 2a^{-1}\left\{\frac{1}{2}\,a + \frac{1}{2} \cdot \frac{1 \cdot 3}{2 \cdot 4}\,a^3 + \frac{1 \cdot 3}{2 \cdot 4} \cdot \frac{1 \cdot 3 \cdot 5}{2 \cdot 4 \cdot 6}\,a^5 + \ldots\right\}.$$

Unless a be small, these series will converge too slowly to be practically useful. More convergent series might be obtained, but Pontécoulant remarks (*Système du Monde*, Tome III. p. 81), that in practice it is more convenient to employ elliptic integrals for the purpose, in the manner we proceed to explain. We have

$$\frac{1}{(a^2 + a'^2 - 2aa'\cos\phi)^{\frac{1}{2}}} = \frac{1}{2}C_0 + C_1\cos\phi + C_2\cos 2\phi + \ldots$$

$$\frac{\cos\phi}{(a^2 + a'^2 - 2aa'\cos\phi)^{\frac{1}{2}}} = \frac{1}{2}C_0\cos\phi + \frac{1}{2}C_1(1 + \cos 2\phi)$$

$$+ \frac{1}{2}C_2(\cos 3\phi + \cos\phi) + \ldots$$

Integrating both sides of these equations with respect to ϕ between the limits 0 and 2π, we obtain

$$\pi C_0 = \int_0^{2\pi} \frac{d\phi}{(a^2 + a'^2 - 2aa'\cos\phi)^{\frac{1}{2}}} = \frac{1}{a}\int_0^{2\pi} \frac{d\phi}{(1 + a^2 - 2a\cos\phi)^{\frac{1}{2}}},$$

$$\pi C_1 = \int_0^{2\pi} \frac{\cos\phi\,d\phi}{(a^2 + a'^2 - 2aa'\cos\phi)^{\frac{1}{2}}} = \frac{1}{a}\int_0^{2\pi} \frac{\cos\phi\,d\phi}{(1 + a^2 - 2a\cos\phi)^{\frac{1}{2}}}.$$

These integrals may be reduced to the standard forms of elliptic functions by assuming

$$\sin(\theta - \phi) = a\sin\theta\ldots\ldots\ldots\ldots\ldots\ldots\ldots(1),$$

whence

$$\tan\theta = \frac{\sin\phi}{\cos\phi - a}\ldots\ldots\ldots\ldots\ldots\ldots\ldots(2).$$

From (1) $\cos(\theta - \phi)\left(1 - \dfrac{d\phi}{d\theta}\right) = \alpha \cos \theta$;

therefore $\dfrac{d\phi}{d\theta} = \dfrac{\cos(\theta - \phi) - \alpha \cos \theta}{\cos(\theta - \phi)}$.

Now

$\cos(\theta - \phi) - \alpha \cos \theta = \cos \theta (\cos \phi - \alpha) + \sin \theta \sin \phi$

$$= \left(\frac{\cos^2 \theta}{\sin \theta} + \sin \theta\right) \sin \phi, \text{ by } (2),$$

$$= \frac{\sin \phi}{\sin \theta}$$

$$= \sqrt{\{(\cos \phi - \alpha)^2 + \sin^2 \phi\}}, \text{ by } (2),$$

$$= \sqrt{(1 + \alpha^2 - 2\alpha \cos \phi)} \quad\ldots\ldots\ldots\ldots(3).$$

Also $\cos(\theta - \phi) = \sqrt{(1 - \alpha^2 \sin^2 \theta)}\ldots\ldots\ldots\ldots\ldots(4)$.

Hence $\dfrac{d\phi}{d\theta} = \sqrt{\left(\dfrac{1 + \alpha^2 - 2\alpha \cos \phi}{1 - \alpha^2 \sin^2 \theta}\right)}$.

Again, from equations (3) and (4)

$$\sqrt{(1 + \alpha^2 - 2\alpha \cos \phi)} = \sqrt{(1 - \alpha^2 \sin^2 \theta)} - \alpha \cos \theta;$$

therefore $1 + \alpha^2 - 2\alpha \cos \phi = 1 - \alpha^2 \sin^2 \theta$

$$+ \alpha^2 \cos^2 \theta - 2\alpha \cos \theta \sqrt{(1 - \alpha^2 \sin^2 \theta)},$$

$$2\alpha \cos \phi = 2\alpha^2 \sin^2 \theta + 2\alpha \cos \theta \sqrt{(1 - \alpha^2 \sin^2 \theta)},$$

or $\cos \phi = \alpha \sin^2 \theta + \cos \theta \sqrt{(1 - \alpha^2 \sin^2 \theta)}$.

Now as ϕ increases from 0 up to 2π, θ also increases from 0 to 2π; hence

$$C_0 = \frac{1}{a\pi} \int_0^{2\pi} \frac{d\phi}{\sqrt{(1 + \alpha^2 - 2\alpha \cos \phi)}}$$

$$= \frac{1}{a\pi} \int_0^{2\pi} \frac{d\theta}{\sqrt{(1 - \alpha^2 \sin^2 \theta)}},$$

$$C_1 = \frac{1}{a\pi} \int_0^{2\pi} \frac{\cos\phi \, d\phi}{\sqrt{(1 + a^2 - 2a \cos\phi)}}.$$

$$= \frac{1}{a\pi} \int_0^{2\pi} \frac{a \sin^2\theta \, d\theta}{\sqrt{(1 - a^2 \sin^2\theta)}} + \frac{1}{a\pi} \int_0^{2\pi} \cos\theta \, d\theta$$

$$= \frac{1}{a^2\pi} \left\{ \int_0^{2\pi} \frac{d\theta}{\sqrt{(1 - a^2 \sin^2\theta)}} - \int_0^{2\pi} \sqrt{(1 - a^2 \sin^2\theta)} \, d\theta \right\}.$$

Hence with the usual notation for elliptic integrals (see Todhunter's *Integral Calculus*, Art. 222),

$$C_0 = \frac{1}{a\pi} F(a, 2\pi) = \frac{4}{a\pi} F\left(a, \frac{\pi}{2}\right),$$

$$C_1 = \frac{1}{a'\pi} \{F(a, 2\pi) - E(a, 2\pi)\}$$

$$= \frac{4}{a'\pi} \left\{ F\left(a, \frac{\pi}{2}\right) - E\left(a, \frac{\pi}{2}\right) \right\}.$$

The numerical values of $F\left(a, \frac{\pi}{2}\right)$ and $E\left(a, \frac{\pi}{2}\right)$ may be found from Legendre's tables of elliptic functions.

54.   *Given C_k and C_{k-1} to obtain C_{k+1}.*

We have

$$(a^2 + a'^2 - 2aa' \cos\phi)^{-\frac{1}{2}} = \frac{1}{2} C_0 + C_1 \cos\phi + \dots + C_k \cos k\phi + \dots;$$

differentiating with respect to ϕ,

$$aa' \sin\phi \, (a^2 + a'^2 - 2aa' \cos\phi)^{-\frac{3}{2}} = C_1 \sin\phi + 2C_2 \sin 2\phi + \dots$$
$$+ kC_k \sin k\phi + \dots;$$

therefore $$aa' \sin\phi \left(\frac{1}{2} C_0 + C_1 \cos\phi + \dots \right)$$

$$= (a^2 + a'^2 - 2aa' \cos\phi) (C_1 \sin\phi + 2C_2 \sin 2\phi + \dots);$$

equating coefficients of $\sin k\phi$,

$$\frac{1}{2}aa'(C_{k-1}-C_{k+1})=k(a^2+a'^2)C_k-aa'\{(k-1)C_{k-1}+(k+1)C_{k+1}\};$$

whence
$$C_{k+1}=\frac{2k}{2k+1}\frac{a^2+a'^2}{aa'}C_k-\frac{2k-1}{2k+1}C_{k-1}.$$

55. *Given C_k and C_{k+1} to obtain D_k.*

As in the last Article, we have

$$aa'\sin\phi\,(a^2+a'^2-2aa'\cos\phi)^{-\frac{3}{2}}=C_1\sin\phi+2C_2\sin 2\phi+\dots$$
$$+kC_k\sin k\phi+\dots;$$

therefore $aa'\sin\phi\left(\frac{1}{2}D_0+D_1\cos\phi+D_2\cos 2\phi+\dots\right)$

$$=C_1\sin\phi+2C_2\sin 2\phi+\dots;$$

equating coefficients of $\sin k\phi$,

$$2kC_k\,aa'(D_{k-1}-D_{k+1})\ \dots\dots\dots\dots (1),$$

writing $k+1$ for k,

$$2(k+1)C_{k+1}=aa'(D_k-D_{k+2})\ \dots\dots\dots\dots (2).$$

Again,

$$(a^2+a'^2-2aa'\cos\phi)^{-\frac{3}{2}}=\frac{1}{2}D_0+D_1\cos\phi+D_2\cos 2\phi+\dots,$$

and

$$(a^2+a'^2-2aa'\cos\phi)^{-\frac{1}{2}}=\frac{1}{2}C_0+C_1\cos\phi+C_2\cos 2\phi+\dots;$$

therefore $\frac{1}{2}C_0+C_1\cos\phi+\dots$

$$=(a^2+a'^2-2aa'\cos\phi)\left(\frac{1}{2}D_0+D_1\cos\phi+\dots\right):$$

equating coefficients of $\cos k\phi$,

$$C_k=(a^2+a'^2)D_k-aa'(D_{k-1}+D_{k+1})\ \dots\dots\dots (3),$$

writing $k+1$ for k,

$$C_{k+1}=(a^2+a'^2)D_{k+1}-aa'(D_k+D_{k+2})\ \dots\dots\dots (4).$$

Eliminating D_{k-1} between (1) and (3),

$$(2k+1)\,C_k = (a^2 + a'^2)\,D_k - 2aa'D_{k+1} \dots\dots (5).$$

Eliminating D_{k+2} between (2) and (4),

$$(2k+1)\,C_{k+1} = -(a^2 + a'^2)\,D_{k+1} + 2aa'D_k. \dots\dots(6).$$

Finally, eliminating D_{k+1} between (5) and (6),

$$(2k+1)\left\{(a^2 + a'^2)\,C_k - 2aa'\,C_{k+1}\right\} = \left\{(a^2 + a'^2)^2 - 4a^2a'^2\right\}D_k,$$

$$\text{or } D_k = \frac{2k+1}{(a^2 - a'^2)^2}\left\{(a^2 + a'^2)\,C_k - 2aa'\,C_{k+1}\right\}.$$

56.   *To calculate the successive differential coefficients of* C_k *and* D_k *with respect to* a *and* a'.

We have

$$(a^2 + a'^2 - 2aa'\cos\phi)^{-\frac{1}{2}} = \frac{1}{2}\,C_0 + C_1\cos\phi + C_2\cos 2\phi + \dots$$

$$+ C_k\cos k\phi + \dots :$$

differentiating with respect to a,

$$-(a - a'\cos\phi)(a^2 + a'^2 - 2aa'\cos\phi)^{-\frac{3}{2}} = \frac{1}{2}\frac{dC_0}{da} + \frac{dC_1}{da}\cos\phi + \dots$$

$$+ \frac{dC_k}{du}\cos k\phi + \dots ;$$

substituting for $(a^2 + a'^2 - 2aa'\cos\phi)^{-\frac{3}{2}}$ its expression in series

$$-(a - a'\cos\phi)\left(\frac{1}{2}\,D_0 + D_1\cos\phi + \dots + D_k\cos k\phi + \dots\right)$$

$$= \frac{1}{2}\frac{dC_0}{da} + \frac{dC_1}{da}\cos\phi + \dots + \frac{dC_k}{da}\cos k\phi + \dots ;$$

equating coefficients of $\cos k\phi$,

$$\frac{dC_k}{da} = -aD_k + \frac{a'}{2}(D_{k-1} + D_{k+1}).$$

By giving to k in succession the values 1, 2, 3, &c., those of $\dfrac{dC_1}{da}$, $\dfrac{dC_2}{da}$, &c. may be found, the right-hand member being calculated by the formula of the preceding Article. By equating the parts independent of ϕ, we obtain

$$\frac{dC_0}{da} = -aD_0 + a'D_1.$$

The value of $\dfrac{dD_k}{da}$ may be found by differentiating the expression for D_k in Art. 55, and substituting for $\dfrac{dC_k}{da}$ and $\dfrac{dC_{k+1}}{da}$ their values as given by the present Article.

The successive differential coefficients of C_k and D_k with respect to a may be obtained from the expressions for $\dfrac{dC_k}{da}$ and $\dfrac{dD_k}{da}$ by simple differentiation and substitution.

57. We might determine in the same way the successive differential coefficients of C_k and D_k with respect to a'; but when those with respect to a have been found, the former may be derived from them, as we proceed to shew. On examining the expansion of $(a^2 + a'^2 - 2aa' \cos \phi)^{-s}$ in Art. 52, it will be seen that A_k is a homogeneous function of a and a' of $-2s$ dimensions. Hence C_k and D_k are homogeneous functions of a and a', the former of -1, the latter of -3 dimensions. It follows that $\dfrac{dC_k}{da}$, $\dfrac{dD_k}{da}$ will be homogeneous functions of -2 and -4 dimensions respectively; and so on. Now by a known property of such functions

$$a \frac{dC_k}{da} + a' \frac{dC_k}{da'} = -C_k,$$

which determines $\dfrac{dC_k}{da'}$:

$$a \frac{d^2 C_k}{da^2} + a' \frac{d^2 C_k}{da\, da'} = -2 \frac{dC_k}{da},$$

which determines $\dfrac{d^2 C_k}{da\, da'}$:

$$a' \frac{d^2 C_k}{da'^2} + a \frac{d^2 C_k}{da\, da'} = -2 \frac{dC_k}{da'},$$

which determines $\dfrac{d^2 C_k}{da'^2}$: and thus all the differential coefficients of C_k may be determined.

In like manner all the successive differential coefficients of D may be calculated.

We are now in a position to simplify the expression for F. We have (Art. 47)

$$F = m' \left\{ \frac{C_0}{2} + \frac{1}{4} \left(a \frac{dC_0}{da} + \frac{a^2}{2} \frac{d^2 C_0}{da^2} \right) e^2 \right.$$

$$+ \frac{1}{4} \left(a' \frac{dC_0}{da'} + a'^2 \frac{d^2 C_0}{da'^2} \right) e'^2$$

$$+ \frac{1}{4} \left(4C_1 + 2a \frac{dC_1}{da} + 2a' \frac{dC_1}{da'} + aa' \frac{d^2 C_1}{da\, da'} \right) ee' \cos (\varpi - \varpi')$$

$$- \frac{1}{8} \left(a^2 D_0 + a \frac{dC_0}{da} \right) \tan^2 i - \frac{1}{8} \left(a'^2 D_0 + a' \frac{dC_0}{da'} \right) \tan^2 i''$$

$$\left. + \frac{1}{4} aa' D_1 \tan i \tan i'' \cos (\Omega - \Omega') + \ldots \right\}.$$

The following proposition will be found useful.

58.　*To shew that* $\dfrac{d^2 C_0}{da\, da'} = - D_1$, *and that* $\dfrac{d^2 C_1}{da\, da'} = - D_0$.

We have

$$\frac{1}{2} C_0 + C_1 \cos \phi + C_2 \cos 2\phi + \ldots = (a' + a'^2 - 2aa' \cos \phi)^{-\frac{1}{2}}$$

therefore $\quad \dfrac{1}{2} \dfrac{dC_0}{da} + \dfrac{dC_1}{da} \cos \phi + \dfrac{dC_2}{da} \cos 2\phi + \ldots$

$$= -(a - a' \cos \phi)(a^2 + a'^2 - 2aa' \cos \phi)^{-\frac{3}{2}};$$

therefore

$$\frac{1}{2}\frac{d^2C_0}{du\,da'} + \frac{d^2C_1}{da\,da'}\cos\phi + \ldots = \cos\phi\,(a^2 + a'^2 - 2aa'\cos\phi)^{-\frac{3}{2}}$$

$$+ 3\,(a - a'\cos\phi)\,(u' - a\cos\phi)\,(a^2 + a'^2 - 2aa'\cos\phi)^{-\frac{5}{2}}$$

$$= \cos\phi\,(a^2 + a'^2 - 2aa'\cos\phi)^{-\frac{3}{2}}$$

$$+ 3\,\{aa'\,(1 + \cos^2\psi) - (a^2 + a'^2)\cos\phi\}\,(a^2 + a'^2 - 2aa'\cos\phi)^{-\frac{5}{2}}$$

$$= \cos\phi\,(a^2 + a'^2 - 2aa'\cos\phi)^{-\frac{3}{2}}$$

$$+ 3\,\{aa'\sin^2\phi - \cos\phi\,(a^2 + a'^2 - 2aa'\cos\phi)\}$$

$$(a^2 + a'^2 - 2aa'\cos\phi)^{-\frac{5}{2}}$$

$$= -2\cos\phi\,(a^2 + a'^2 - 2aa'\cos\phi)^{-\frac{3}{2}}$$

$$+ 3aa'\sin^2\phi\,(a^2 + a'^2 - 2aa'\cos\phi)^{-\frac{5}{2}}.$$

Now $(a^2 + a'^2 - 2aa'\cos\phi)^{-\frac{3}{2}} = \dfrac{1}{2}D_0 + D_1\cos\phi + \ldots;$

differentiating with respect to ϕ

$$3aa'\sin\phi\,(a^2 + a'^2 - 2aa'\cos\phi)^{-\frac{5}{2}}$$

$$= D_1\sin\phi + 2D_2\sin 2\phi + \ldots;$$

therefore $\dfrac{1}{2}\dfrac{d^2C_0}{da\,da'} + \dfrac{d^2C_1}{da\,da'}\cos\phi + \ldots$

$$= -2\cos\phi\left(\frac{1}{2}D_0 + D_1\cos\phi + \ldots\right)$$

$$+ \sin\phi\,(D_1\sin\phi + 2D_2\sin 2\phi + \ldots),$$

whence, equating the parts independent of ϕ, and also the coefficients of $\cos\phi$

$$\frac{1}{2}\frac{d^2C_0}{da\,da'} = -D_1 + \frac{D_1}{2} = -\frac{D_1}{2},$$

or

$$\frac{d^2C_0}{da\,da'} = -D_1;$$

and

$$\frac{d^2C_1}{da\,da'} = -D_0.$$

59. Since $\dfrac{dC_0}{da}$ is a homogeneous function of a and a' of -2 dimensions,

$$a\,\frac{d^2C_0}{da^2} + a'\,\frac{d^2C_0}{da\,da'} = -2\,\frac{dC_0}{da};$$

therefore

$$a\,\frac{dC_0}{da} + \frac{a^2}{2}\,\frac{d^2C_0}{da^2} = -\frac{aa'}{2}\,\frac{d^2C_0}{da\,da'}$$

$$= \frac{1}{2}\,aa'\,D_1.$$

Similarly, $$a'\,\frac{dC_0}{da'} + \frac{a'^2}{2}\,\frac{d^2C_0}{da'^2} = \frac{1}{2}\,aa'\,D_1.$$

Hence the coefficients of e^2 and e'^2 in the expression for F are each equal to $\dfrac{1}{8}\,aa'\,D_1$.

Again, since C_1 is a homogeneous function of a and a' of -1 dimensions,

$$a\,\frac{dC_1}{da} + a'\,\frac{dC_1}{da'} = -C_1;$$

hence the coefficient of $ee'\cos(\varpi - \varpi')$

$$= \frac{1}{4}\,(2C_1 - aa'\,D_0);$$

but (Art. 55) $$2kC_k = aa'\,(D_{k-1} - D_{k+1});$$

therefore, making $$k = 1,$$

$$2C_1 = aa'\,(D_0 - D_2);$$

hence the coefficient of $ee'\cos(\varpi - \varpi')$

$$= -\frac{1}{4}\,aa'\,D_2.$$

Again, (Art. 56)

$$\frac{dC_0}{da} = -aD_0 + a^2 D_1;$$

therefore
$$a^2 D_0 + a\,\frac{dC_0}{da} = aa' D_1.$$

Similarly,
$$a'^2 D_0 + a'\,\frac{dC_0}{da'} = aa' D_1.$$

Hence the coefficients of $\tan^2 i$ and $\tan^2 i'$ are each equal
to $-\dfrac{1}{8}\,aa' D_1$.

Finally, the expression for F becomes

$$F = m'\left\{ \frac{C_0}{2} + \frac{1}{8}\,aa' D_1\,(e^2 + e'^2) - \frac{1}{4}\,aa' D_2\,ee'\cos(\varpi - \varpi')\right.$$

$$-\frac{1}{8}\,aa' D_1\,(\tan^2 i + \tan^2 i') + \frac{1}{4}\,aa' D_1 \tan i \tan i' \cos(\Omega - \Omega')$$

$$\left. + \ldots \right\}.$$

CHAPTER IV.

SECULAR VARIATIONS OF THE ELEMENTS OF THE ORBIT. STABILITY OF THE PLANETARY SYSTEM.

60. WE have seen in the preceding Chapter, that the disturbing function, when developed, consists of two parts; the one independent of the time explicitly, the other involving it under a periodical form: we shall consider separately the effects of these two parts. In the present Chapter our attention will be directed to the first or non-periodical part of R, which we have denoted by F. The inequalities thus produced in the elements of the orbit are termed *secular*, in consequence of their very slow variation.

61. By differentiating the expression for F in Art. 59, with respect to the elements, we obtain

$$\frac{dF}{d\epsilon} = 0,$$

$$\frac{dF}{d\varpi} = \frac{m'}{4} aa' D_2 ee' \sin (\varpi - \varpi'),$$

$$\frac{dF}{de} = \frac{m'}{4} aa' D_1 e - \frac{m'}{4} aa' D_2 e' \cos (\varpi - \varpi'),$$

$$\frac{dF}{d\Omega} = - \frac{m'}{4} aa' D_1 \tan i \tan i'' \sin (\Omega - \Omega'),$$

$$\frac{dF}{di} = -\frac{m'}{4} \cdot aa' D_1 \tan i + \frac{m'}{4} aa' D_1 \tan i' \cos(\Omega - \Omega'),$$

$$\frac{dF}{da} = \text{an expression similar to } F.$$

Substituting these in the formulæ of Art. 39, and neglecting powers of e, e', $\tan i$, $\tan i'$ higher than the second, we have

$$\frac{da}{dt} = 0,$$

$$\frac{de}{dt} = -\frac{m'na^2a'}{4\mu} D_2 e' \sin(\varpi - \varpi'),$$

$$\frac{d\varpi}{dt} = \frac{m'na^2a'}{4\mu} \{ D_1 e - D_2 o' \cos(\varpi - \varpi') \},$$

$$\frac{di}{dt} = \frac{m'na^2a'}{4\mu} D_1 \tan i' \sin(\Omega - \Omega'),$$

$$\tan i \frac{d\Omega}{dt} = -\frac{m'na^2a'}{4\mu} D_1 \{ \tan i - \tan i' \cos(\Omega - \Omega') \},$$

$$\frac{d\epsilon}{dt} = A + A_1 (e^2 - \tan^2 i) + A_2 (e'^2 - \tan^2 i')$$

$$+ A_3 ee' \cos(\varpi - \varpi') + A_4 \tan i \tan i' \cos(\Omega - \Omega'),$$

where in the last expression, A, A_1, &c., have been written to denote certain functions of a and a'.

62. *To calculate approximately the secular variations of the elements of a planet's orbit, in a given time.*

Let a_0, e_0, ϖ_0, &c., be the values of the elements at some given epoch; $a_0 + \delta a^*$, $e_0 + \delta e$, $\varpi_0 + \delta\varpi$, &c., their values after an interval t: then δa, δe, $\delta\varpi$, &c., are the required variations. By Maclaurin's Theorem,

* It will be shewn in Art. 64 that δa is always zero,

$$\delta e = \left(\frac{de}{dt}\right)_0 t + \left(\frac{d^2e}{dt^2}\right)_0 \frac{t^2}{\underline{2}} + \dots$$

$$\delta \varpi = \left(\frac{d\varpi}{dt}\right)_0 t + \left(\frac{d^2\varpi}{dt^2}\right)_0 \frac{t^2}{\underline{|2}} + \dots$$

$$\dots = \dots$$

which may be carried to any required degree of accuracy, but in practice the first two terms will generally be sufficient.

We have supposed the variations of the elements required at a time t *after* the epoch; if they be required at a time t *before* the epoch, we have only to change the sign of t in the above.

We may remark that $\left(\frac{de}{dt}\right)_0$, $\left(\frac{d\varpi}{dt}\right)_0$, &c. are of the order of the disturbing force, since they involve the first power of m': $\left(\frac{d^2e}{dt^2}\right)_0$, $\left(\frac{d^2\varpi}{dt^2}\right)_0$, &c. are of the second order; for, since the expressions for $\frac{de}{dt}$, $\frac{d\varpi}{dt}$, &c., involve elements, their differential coefficients will involve the differential coefficients of those elements, and thus, by substitution, m'^2 will be introduced: similarly $\left(\frac{d^3e}{dt^3}\right)_0$, $\left(\frac{d^3\varpi}{dt^3}\right)_0$, &c., are of the third order, and so on.

In the short period of one year all terms after the first may be neglected, so that putting $t = 1$, we have

$$\delta e = \left(\frac{de}{dt}\right)_0, \text{ &c.}$$

Hence the coefficient of t in the above formulæ is called the *annual variation*.

63. Since the elements of the planetary orbits are continually changing, it will be interesting to shew that the

dimensions of these orbits, and their inclinations to the ecliptic, nevertheless fluctuate between very narrow limits. This constitutes what is termed the Stability of the Planetary System : in order to establish it, it will be necessary to prove the stability (i) of the mean distances, (ii) of the exceutricities, (iii) of the inclinations.

64. *To prove the stability of the mean distances of the planets from the Sun, and of their mean motions.*

By Art. 61 $\dfrac{da}{dt} = 0$, so that a is constant. Now it will be shewn in a subsequent Chapter (see Art. 91), that to the first order of the disturbing force, the periodical terms of R can produce only periodical variations; consequently, to this order, the mean distance is susceptible of no permanent change*. The same is true of the mean motion n, since it $= \dfrac{\mu^{\frac{1}{2}}}{a^{\frac{3}{2}}}$, and μ does not alter. We are hereby assured of the impossibility of any of the bodies of our system ever leaving it in consequence of the disturbances it may experience from the other bodies; and this secures the general permanence of the whole, by keeping the mean distances and periodic times perpetually fluctuating between certain limits (very restricted ones) which they can never exceed or fall short of.

This result may easily be extended to all orders of the excentricities and inclinations : for since $nt + \epsilon$ always occurs in R as one symbol, ϵ cannot occur in F because t does not, so that $\dfrac{dF}{d\epsilon}$, and therefore $\dfrac{da}{dt}$ is zero.

---

* This result is also true when the square of the disturbing force is included : for the demonstration the reader is referred to Pontécoulant's *Système du Monde*, Tome I. p. 395 (2nd edit.).

65. *To prove the stability of the excentricities of the planetary orbits.*

We will first consider the case of two planets only. By Art. 61,

$$\frac{de}{dt} = -\frac{m'na^2a'}{4\mu} D_2 e' \sin(\varpi - \varpi').$$

Similarly, $\dfrac{de'}{dt} = -\dfrac{mn'a'^2a}{4\mu} D_2'e \sin(\varpi' - \varpi).$

Now since D_2 is the coefficient of $\cos 2\phi$ in the development of

$$(a^2 + a'^2 - 2aa' \cos \phi)^{-\frac{3}{2}},$$

an expression in which a and a' are similarly involved, it follows that

$$D_2' = D_2.$$

Hence, multiplying the above equations by $\dfrac{m}{na} e$, $\dfrac{m'}{n'a'} e'$, respectively, and adding, we have

$$\frac{m}{na} e \frac{de}{dt} + \frac{m'}{n'a'} e' \frac{de'}{dt} = 0;$$

therefore, since a experiences no secular variation,

$$\frac{m}{na} e^2 + \frac{m'}{n'a'} e'^2 = C.$$

A similar equation holds for any number of planets. Replacing for convenience $\dfrac{aa'D_2}{4\mu}$ by (a, a'), we have

$$\frac{de}{dt} = - m'na\,(a, a')\,e' \sin(\varpi - \varpi')$$

$$- m''na\,(a, a'')\,e'' \sin(\varpi - \varpi'') - \ldots$$

$$\frac{de'}{dt} = - mn'a'\,(a', a)\,e \sin(\varpi' - \varpi)$$

$$- m''n'a'\,(a', a'')\,e'' \sin(\varpi' - \varpi'') - \ldots$$

$$\frac{de''}{dt} = -mn''a''\,(a'',\,a)\,e\sin\,(\varpi''-\varpi)$$

$$-m'n''a''\,(a'',\,a')\,e'\sin\,(\varpi''-\varpi')-\ldots$$

$$\ldots\ldots = \ldots\ldots\ldots\ldots$$

Since $D_2' = D_2$, it follows that $(a,\,a') = (a',\,a)$: hence multiplying these equations by $\frac{m}{na}\,e$, $\frac{m'}{n'a'}\,e'$, &c., and adding, we obtain

$$\frac{m}{na}\,e\,\frac{de}{dt} + \frac{m'}{n'a'}\,e'\,\frac{de'}{dt} + \frac{m''}{n''a''}\,e''\,\frac{de''}{dt} + \ldots = 0;$$

whence by integration

$$\Sigma\left(\frac{m}{na}\,e^2\right) = \text{const.}$$

Since $na = \sqrt{\dfrac{\mu}{a}}$, this may be written

$$\Sigma\,(m\,\sqrt{a e^2}) = C.$$

Now observation shews that all the planets revolve round the Sun in the same direction, so that the mean motions n, n', n'', &c. are of uniform sign. Hence all the terms of the left-hand member of the above equation are positive.

We learn also from observation that the excentricities of the planetary orbits are at present very small indeed, and in the case of the Asteroids, the masses are very small. Hence the constant must be small. Since, then, all the terms of the first side of the equation are positive, and their sum always equals a small constant, it follows that every term is small; and therefore, except in the case of planets of small mass, such as Mercury, Mars, Juno, &c., that the excentricities must remain permanently small.

The stability of the excentricities, however, is not confined to the larger planets: we shall give another proof of this important theorem and that of Art. 66 in the following Chapter.

66. *To prove the stability of the inclinations of the planes of the planetary orbits.*

By Art. 61 $\dfrac{di}{dt} = \dfrac{m'na^2a'}{4\mu} D_1 \tan i' \sin (\Omega - \Omega')$.

Similarly, $\dfrac{di''}{dt} = \dfrac{mn'a'^2a}{4\mu} D_1' \tan i \sin (\Omega' - \Omega)$.

As in the preceding Article, it may be shewn that $D_1' = D_1$. Hence, multiplying the above equations by

$$\frac{m}{na} \tan i, \quad \frac{m'}{n'a'} \tan i'',$$

respectively, and adding, we have

$$\frac{m}{na} \tan i \cdot \frac{di}{dt} + \frac{m'}{n'a'} \tan i'' \frac{di''}{dt} = 0,$$

or to the same order of approximation,

$$\frac{m}{na} \tan i \cdot \frac{d(\tan i)}{dt} + \frac{m'}{n'a'} \tan i'' \frac{d(\tan i'')}{dt} = 0;$$

therefore $\qquad \dfrac{m}{na} \tan^2 i + \dfrac{m'}{n'a'} \tan^2 i'' = \text{const.},$

or, since $\qquad\qquad na = \sqrt{\dfrac{\mu}{a}},$

$$m \sqrt{a} \tan^2 i + m' \sqrt{a'} \tan^2 i'' = C.$$

A similar equation would (as in the case of the excentricities) be true for any number of planets. Now the inclinations of the planetary orbits to the ecliptic are at present very small; hence, if we take for our fixed plane of reference a plane coinciding with the present position of the ecliptic, and except the case of planets of small mass, it follows, as in

Art. 65, that their inclinations to this plane must always remain very small.

67.   It should be noticed that the assumption in the two preceding Articles, that μ is the same for all the planets, is equivalent to neglecting the square of the disturbing force: for, let μ refer to the planet m and μ' to m'; then, if M denote the Sun's mass, $\mu = M + m$, $\mu' = M + m'$, so that μ' differs from μ by a quantity of the order of the disturbing force: since, then, the expressions in which μ and μ' occur are themselves of the first order, it follows that the error introduced by supposing μ and μ' equal is of the second order.

68.   The results of Arts. 65 and 66 may also be obtained directly from the equations of motion.   We have (Arts. 20 and 15)

$$\frac{dh}{dt} = \frac{dR}{d\theta} = \frac{dR}{d\epsilon} + \frac{dR}{d\varpi},$$

or, replacing R by F, since we have seen that $\dfrac{dF}{d\epsilon} = 0$,

$$\frac{dh}{dt} = \frac{dF}{d\varpi}.$$

Multiplying both sides of this equation by m, forming similar equations for each planet of the system, and adding, we have

$$\Sigma \left(m \frac{dh}{dt} \right) = \Sigma \left(m \frac{dF}{d\varpi} \right).$$

Now on referring to the expression for F in Art. 59, since ϖ occurs only in the term $-\dfrac{m'}{4} aa' D_2 ee' \cos (\varpi - \varpi')$, it is easily seen that $\Sigma \left(m \dfrac{dF}{d\varpi} \right) = 0$; hence our equation becomes

$$\Sigma \left(m \frac{dh}{dt} \right) = 0,$$

whence by integration

$$\Sigma (mh) = \text{const.};$$

or since $\qquad h^2 = \mu a (1 - e^2) = (M + m) a (1 - e^2),$

$$\Sigma \left\{ m \sqrt{(Ma)} \left(1 + \frac{1}{2} \frac{m}{M} + \dots \right) \left(1 - \frac{e^2}{2} + \dots \right) \right\} = \text{const.}$$

Since a is constant as regards secular variations, and our approximation extends only to the second order of the excentricities, we have to the first order of the disturbing force

$$\Sigma (m \sqrt{ae^2}) = C,$$

the equation of Art. 65.

Again, by referring the motion to the fixed plane of reference, we obtain

$$\frac{d}{dt} (h \cos i) = \frac{dR}{d\theta_1} = \frac{dR}{d\epsilon} + \frac{dR}{d\varpi} + \frac{dR}{d\Omega} \quad \text{(Art. 16),}$$

and considering the whole planetary system, we have as before

$$\Sigma \left\{ m \frac{d}{dt} (h \cos i) \right\} = \Sigma \left\{ m \left(\frac{dF}{d\varpi} + \frac{dF}{d\Omega} \right) \right\}.$$

Now on referring to the expression for F, it is easily seen from the forms under which ϖ and Ω occur, that

$$\Sigma \left(m \frac{dF}{d\varpi} \right) = 0, \quad \Sigma \left(m \frac{dF}{d\Omega} \right) = 0;$$

hence our equation becomes

$$\Sigma \left\{ m \frac{d}{dt} (h \cos i) \right\} = 0,$$

whence by integration

$$\Sigma (mh \cos i) = \text{const.}^*,$$

---

* It should be borne in mind that this and the equation $\Sigma (mh) = C$, which are those which would be obtained by conservation of areas were it lawful to

or $$\Sigma \left\{ mh \left(1 - 2 \sin^2 \frac{i}{2} \right) \right\} = \text{const.},$$

which, from above, may be written

$$\Sigma \left(mh \sin^2 \frac{i}{2} \right) = \text{const.},$$

and proceeding as before, we obtain to the second order of the excentricities and inclinations, and the first of the disturbing force

$$\Sigma \left(m \sqrt{a} \sin^2 \frac{i}{2} \right) = \text{const.},$$

or, since to this order of approximation, the inclinations, their sines and tangents are equal, this may be written

$$\Sigma \left(m \sqrt{a} \tan^2 i \right) = C,$$

the equation of Art. 66.

We may remark that the equation

$$\Sigma \left(mh \cos i \right) = \text{const.}$$

is of itself sufficient to establish the stability both of the excentricities and inclinations. For, proceeding as before, we obtain

$$\Sigma \left\{ m \sqrt{(Ma)} \left(1 + \frac{1}{2} \frac{m}{M} + \dots \right) \left(1 - \frac{e^2}{2} + \dots \right) \left(1 - \frac{i^2}{2} + \dots \right) \right\} = \text{const.},$$

which, to the second order of the excentricities and inclinations and the first of the disturbing force, gives

$$\Sigma \left(m \sqrt{a} e^2 \right) + \Sigma \left(m \sqrt{a} i^2 \right) = C,$$

or to the same order of approximation

$$\Sigma \left(m \sqrt{a} e^2 \right) + \Sigma \left(m \sqrt{a} \tan^2 i \right) = C.$$

Since we know from observation that all the planets revolve round the Sun in the same direction, all the radicals

assume the principle, are true only as regards secular variations. To assume their actual truth would be to neglect terms in R due to the Sun's motion, of the first order of the disturbing force.

in this equation must be taken with the same sign. Also, since the excentricities and inclinations are at present very small, the constant must be small. Hence it follows, as in Arts. 65 and 66, that the excentricities and inclinations must always remain very small.

69.   From the preceding analysis we draw the following remarkable conclusion : *The fact that the planets revolve about the Sun in the same direction, ensures the stability of the planetary system.*   The converse of this would not necessarily be true, as we shall see in Art. 75 : the numerical relations of the dimensions and positions of the orbits of the planets, might be such as to ensure stability, although they revolved in opposite directions.   But the above is independent of particular numerical relations.

70.   The results at which we have arrived with regard to the stability of the planetary system are of especial interest. In consequence of the changes in the elements it might have been supposed that the orbits would ultimately undergo such alterations in their dimensions as to bring the planets into collision or hurry them into boundless space.   Or even if no such violent catastrophe occurred, a derangement of the seasons might seriously have interfered with the physical comfort of man*.   But our analysis shews, (and the results are confirmed when the approximation is carried further,) that in so far as the mutual attractions of the Sun and planets are concerned, the dimensions and position of the orbits will for ages remain nearly the same as they are at present, i.e. nearly circular in form, and but little inclined to each other, thus affording a beautiful illustration of Gen. viii. 22 : "While the earth remaineth, seed-time and harvest, and cold and heat, and summer and winter, and day and night shall not cease."

* See Herschel's *Outlines of Astronomy.*

CHAPTER V.

SECULAR VARIATIONS OF THE ELEMENTS CONTINUED.
INTEGRATION OF THE DIFFERENTIAL EQUATIONS.

71. In Art. 62 we have given a method of calculating the secular variations sufficiently accurate for the practical purposes of astronomy, but in order to understand their real nature, and thus to examine more fully into the important question of the stability of the excentricities and inclinations, it is necessary to proceed to the actual integration of the equations of Art. 61. This we are enabled to do by a method due to Lagrange.

72. *To integrate the equations for the excentricity and longitude of perihelion.*

We have (Art. 61) for the planet m

$$\frac{de}{dt} = -\frac{m'na^2a'}{4\mu} D_2 e' \sin(\varpi - \varpi'),$$

$$e\frac{d\varpi}{dt} = \frac{m'na^2a'}{4\mu} \{D_1 e - D_2 e' \cos(\varpi - \varpi')\};$$

with similar equations for the planet m'.

We shall be able to reduce these to a system of linear differential equations if we assume

$$u = e \sin \varpi, \qquad v = e \cos \varpi,$$
$$u' = e' \sin \varpi', \qquad v' = e' \cos \varpi';$$

therefore $\dfrac{du}{dt} = e \cos \varpi \dfrac{d\varpi}{dt} + \sin \varpi \dfrac{de}{dt}$.

Substituting the values of $\dfrac{d\varpi}{dt}$ and $\dfrac{de}{dt}$, and writing α for $\dfrac{m' n a^2 a'}{4 \mu}$, we have

$$\frac{du}{dt} = \alpha \left(D_1 e \cos \varpi - D_2 e' \cos \varpi' \right)$$

$$= \alpha \left(D_1 v - D_2 v' \right).$$

Similarly, $\dfrac{dv}{dt} = \alpha \left(D_2 u' - D_1 u \right).$

In like manner for the planet m', writing α' for $\dfrac{m n' a'^2 a}{4 \mu}$, we have

$$\frac{du'}{dt} = \alpha' \left(D_1 v' - D_2 v \right),$$

$$\frac{dv'}{dt} = \alpha' \left(D_2 u - D_1 u' \right).$$

The forms of these equations suggest the following particular integrals :

$$u = M \sin (gt + \gamma), \qquad v = M \cos (gt + \gamma),$$
$$u' = M' \sin (gt + \gamma), \qquad v' = M' \cos (gt + \gamma).$$

Substituting these in the differential equations, we obtain from either of the first two

$$gM = \alpha \left(D_1 M - D_2 M' \right),$$

and from either of the last two

$$gM' = \alpha' \left(D_1 M' - D_2 M \right);$$

eliminating the ratio $M : M'$

$$(g - aD_1)(g - a'D_1) = aa'D_2^2,$$

or $g^2 - (a + a') D_1 g + aa' (D_1^2 - D_2^2) = 0 :$

and the roots of this equation will be real and unequal, real and equal, or impossible, according as

$$(a + a')^2 D_1^2 - 4aa' (D_1^2 - D_2^2)$$

is positive, zero, or negative. Now

$$(a + a')^2 D_1^2 - 4aa'(D_1^2 - D_2^2) = (a - a')^2 D_1^2 + 4aa'D_2^2,$$

a positive quantity, since n, n' and therefore a, a' are of like sign. Hence the values of g will be real and unequal: denote them by g_1, g_2, and let γ_1, γ_2; M_1, M_2; M_1', M_2'; be the corresponding values of γ, M, M' respectively. Then the complete solution of the differential equations will be

$$u = M_1 \sin (g_1 t + \gamma_1) + M_2 \sin (g_2 t + \gamma_2),$$
$$v = M_1 \cos (g_1 t + \gamma_1) + M_2 \cos (g_2 t + \gamma_2),$$
$$u' = M_1' \sin (g_1 t + \gamma_1) + M_2' \sin (g_2 t + \gamma_2),$$
$$v' = M_1' \cos (g_1 t + \gamma_1) + M_2' \cos (g_2 t + \gamma_2).$$

Of the constants in these equations, four are arbitrary and must be determined from observation. We have

$$e^2 = u^2 + v^2 = M_1^2 + M_2^2 + 2M_1 M_2 \cos \{(g_1 - g_2) t + \gamma_1 - \gamma_2\},$$

$$\tan \varpi = \frac{u}{v} = \frac{M_1 \sin (g_1 t + \gamma_1) + M_2 \sin (g_2 t + \gamma_2)}{M_1 \cos (g_1 t + \gamma_1) + M_2 \cos (g_2 t + \gamma_2)} :$$

with similar equations for e' and ϖ'.

73. Had we considered a system of several planets, we should have obtained by a like process

$$e^2 = M_1^2 + M_2^2 + M_3^2 + \ldots + 2M_1 M_2 \cos \{(g_1 - g_2) t + \gamma_1 - \gamma_2\}$$
$$+ 2M_1 M_3 \cos \{(g_1 - g_3) t + \gamma_1 - \gamma_3\} + \ldots$$

$$\tan \varpi = \frac{M_1 \sin (g_1 t + \gamma_1) + M_2 \sin (g_2 t + \gamma_2) + M_3 \sin (g_3 t + \gamma_3) + \ldots}{M_1 \cos (g_1 t + \gamma_1) + M_2 \cos (g_2 t + \gamma_2) + M_3 \cos (g_3 t + \gamma_3) + \ldots},$$

with similar equations for each of the other planets.

74. We may hence infer the stability of the excentricities. From the form of the expression for e it appears
that e^2 cannot be greater than

$$M_1^2 + M_2^2 + M_3^2 + \ldots + 2M_1M_2 + 2M_1M_3 + \ldots,$$

and therefore that e cannot exceed

$$M_1 + M_2 + M_3 + \ldots,$$

these quantities being all taken with the same sign.

Thus, by determining the numerical values of M_1, M_2,
&c., for any particular planet, we may assign an actual limit
which its excentricity can never exceed. For the principal
planets, M_1, M_2, &c., are found to be exceedingly small, so
that the excentricity must always remain very small.

If we confine our attention to two planets, we see
from the expression for e in Art. 72, that the excentricity
fluctuates between the limits $M_1 + M_2$ and $M_1 \sim M_2$. The
period of these changes $= \dfrac{2\pi}{g_1 \sim g_2}$, and is the same for each
planet: it appears also from the equation

$$m \sqrt{ae^2} + m' \sqrt{a'e'^2} = C,$$

obtained in Art. 65, that the maximum of each excentricity
takes place at the time of the minimum of the other.

As an illustration, take the case of Jupiter and Saturn.
Sir John Herschel finds that

$$g_1 = 21''{\cdot}9905, \quad g_2 = 3''{\cdot}5851;$$
$$M_1 = -{\cdot}01715, \quad M_2 = {\cdot}04321, \text{ for Jupiter};$$
$$M_1' = {\cdot}04877, \quad M_2' = {\cdot}03532, \text{ for Saturn}:$$

the year 1700 being taken as the epoch[*]. Thus we obtain
for the greatest and least excentricities that Jupiter's orbit
can attain, ·06036 and ·02606 respectively, and for those of

[*] Article *Physical Astronomy* in the *Encyclopædia Metropolitana*.

Saturn, ·08409 and ·01345, quantities exceedingly minute; while the period of these changes amounts to no less than 70414 years.

75. It appears from the preceding Article that the stability of the excentricities is a consequence of the periodical form of the solution of the differential equations, a result which depends upon the fact that g_1 and g_2 are real and unequal. Now we have seen that in order that this may be the case, it is only necessary that

$$(\alpha + \alpha')^2 D_1^2 - 4\alpha\alpha' (D_1^2 - D_2^2)$$

shall be positive, a condition which might be satisfied if the signs of n, n', and therefore of α, α' were different. In this case, then, the stability would still subsist. Let us, however, consider what would be the effect of equal or impossible roots to the quadratic from which g is found. In the former case a term would be introduced into u, u', v, and v' proportional to the time, and in the latter the periodical terms would be replaced by exponentials. Consequently the excentricities would increase indefinitely with the time, and the stability would no longer subsist.

76. We now proceed to examine the expression which has been obtained in Art. 72, for the longitude of perihelion, viz.

$$\tan \varpi = \frac{M_1 \sin (g_1 t + \gamma_1) + M_2 \sin (g_2 t + \gamma_2)}{M_1 \cos (g_1 t + \gamma_1) + M_2 \cos (g_2 t + \gamma_2)};$$

$$\therefore \frac{d\varpi}{dt} = \frac{g_1 M_1^2 + g_2 M_2^2 + (g_1 + g_2) M_1 M_2 \cos \{(g_1 - g_2) t + \gamma_1 - \gamma_2\}}{M_1^2 + M_2^2 + 2 M_1 M_2 \cos \{(g_1 - g_2) t + \gamma_1 - \gamma_2\}}.$$

The maxima and minima values of ϖ, if such exist, will be found by equating $\dfrac{d\varpi}{dt}$ to zero. Thus

$$\cos \{(g_1 - g_2) t + \gamma_1 - \gamma_2\} = -\frac{g_1 M_1^2 + g_2 M_2^2}{(g_1 + g_2) M_1 M_2}.$$

If this (disregarding sign) be not greater than unity, the perihelion will oscillate, the period of a complete oscillation being the same as that of the excentricities, viz. $\dfrac{2\pi}{g_1 \sim g_2}$; but if, as is the case with Jupiter and Saturn, this be greater than unity, the longitude of perihelion has no maximum or minimum, and the perihelion moves constantly in one direction.

Again,

$$\frac{d\varpi}{dt} = \frac{1}{2} \frac{(g_1 - g_2)(M_1^2 - M_2^2)}{M_1^2 + M_2^2 + 2M_1 M_2 \cos\{(g_1 - g_2)t + \gamma_1 - \gamma_2\}} + \frac{1}{2}(g_1 + g_2)$$

$$= \frac{(g_1 - g_2)(M_1^2 - M_2^2)}{2e^2} + \frac{1}{2}(g_1 + g_2).$$

Hence when e is a maximum or minimum, $\dfrac{d\varpi}{dt}$ will be either a maximum or minimum, and the apsidal line will be moving most rapidly or most slowly, different cases occurring according to the signs and magnitudes of the quantities involved.

77. *When the apsidal line oscillates, to find the extent and periods of its oscillations.*

We have (Art. 72)

$$\tan \varpi = \frac{M_1 \sin (g_1 t + \gamma_1) + M_2 \sin (g_2 t + \gamma_2)}{M_1 \cos (g_1 t + \gamma_1) + M_2 \cos (g_2 t + \gamma_2)};$$

therefore
$$\tan (\varpi - g_1 t - \gamma_1) = \frac{\tan \varpi - \tan (g_1 t + \gamma_1)}{1 + \tan \varpi \tan (g_1 t + \gamma_1)}$$

$$= \frac{-M_2 \sin \psi}{M_1 + M_2 \cos \psi},$$

if $\psi = (g_1 - g_2)t + \gamma_1 - \gamma_2$.

Also by the last Article, if τ be the least positive angle whose cosine is $\dfrac{g_1 M_1^2 + g_2 M_2^2}{(g_1 + g_2) M_1 M_2}$,

$$e^2 \frac{d\varpi}{dt} = (g_1 + g_2) M_1 M_2 (\cos\psi + \cos\tau).$$

Different cases will occur according to the signs of M_1, M_2, &c. Suppose M_1, M_2 of like sign, g_1 and g_2 positive, and g_1 greater than g_2. Then ψ increases as t increases, and $\dfrac{d\varpi}{dt}$ will be negative, or the apsidal line will regrede, while $\cos\psi + \cos\tau$ is negative, i.e. so long as ψ is between $(2n-1)\pi - \tau$ and $(2n-1)\pi + \tau$: $\dfrac{d\varpi}{dt}$ will be positive, or the apsidal line will progrede, while ψ is between $(2n-1)\pi + \tau$ and $(2n+1)\pi - \tau$.

To find the angle through which the apsidal line regredes and the period of the regression. Let t', t'' be the values of t, ϖ', ϖ'' the values of ϖ corresponding to the values $(2n-1)\pi - \tau$ and $(2n-1)\pi + \tau$ of ψ: then

$$(g_1 - g_2) t' + \gamma_1 - \gamma_2 = (2n-1)\pi - \tau,$$
$$(g_1 - g_2) t'' + \gamma_1 - \gamma_2 = (2n-1)\pi + \tau,$$
$$\tan(\varpi' - g_1 t' - \gamma_1) = \frac{-M_2 \sin\tau}{M_1 - M_2 \cos\tau},$$
$$\tan(\varpi'' - g_1 t'' - \gamma_1) = \frac{M_2 \sin\tau}{M_1 - M_2 \cos\tau}.$$

From these equations the values of t', t'', ϖ', ϖ'', may be found, and thus $\varpi' - \varpi''$ the amount of regression will be known. The period of regression

$$= t'' - t' = \frac{2\tau}{g_1 - g_2}.$$

In like manner the amount and period of the progression may be obtained. The latter will be found to be $\dfrac{2(\pi - \tau)}{g_1 - g_2}$.

The period of a complete oscillation will be the sum of the periods of the regression and progression, that is $\dfrac{2\pi}{g_1 - g_2}$, which agrees with the preceding Article.

78. The motion of the centre of the instantaneous ellipse in consequence of the secular variations of e and ϖ may be exhibited geometrically as follows.

We have, by Art. 72,

$$e \cos \varpi = M_1 \cos (g_1 t + \gamma_1) + M_2 \cos (g_2 t + \gamma_2),$$
$$e \sin \varpi = M_1 \sin (g_1 t + \gamma_1) + M_2 \sin (g_2 t + \gamma_2).$$

Let a circle be described in the plane of the orbit with its centre S coinciding with that of the Sun, and its radius equal to $M_1 a$, where a is the mean distance. Let a point P describe this circle uniformly with a velocity g_1, starting from

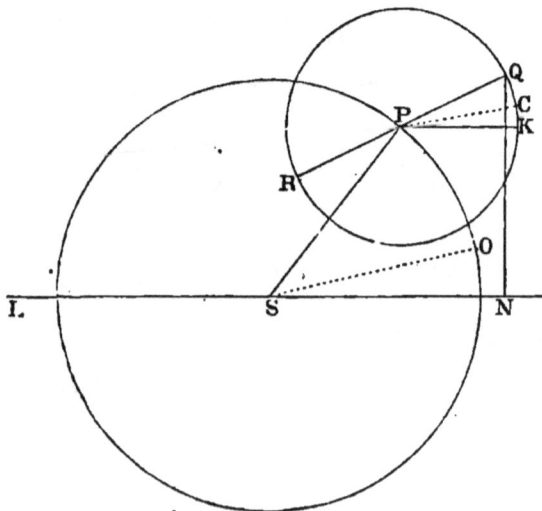

O. Again, with centre P and radius equal to $M_2 a$ let another circle be described, and let a point Q describe this circle uniformly with a velocity g_2, starting from C. Let SL be the line from which longitudes are reckoned, and draw PK parallel to it: then if the angle OSN be equal to γ_1, and CPK to γ_2, the angle PSN will be equal to $g_1 t + \gamma_1$, and QPK to $g_2 t + \gamma_2$. Produce QP to meet the circle again in R, and draw QN perpendicular to SN. Then, supposing M_1 and M_2 to be both positive, we have

$$SN = SP \cos PSN + PQ \cos QPK$$

$$= M_1 a \cos (g_1 t + \gamma_1) + M_2 a \cos (g_2 t + \gamma_2)$$

$$= ae \cos \varpi.$$

Similarly, it may be shewn that

$$QN = ae \sin \varpi.$$

Hence, the apse being supposed to move from L in the direction contrary to that of the hands of a watch, Q will be the centre of the instantaneous ellipse.

If M_1 be positive and M_2 negative, it may be shewn in like manner that the centre of the ellipse will be R. If M_1, M_2 be both negative, join QS and produce it to Q' so that $SQ' = SQ$: then the centre of the ellipse will be Q'.

A similar construction will of course apply for the motion of the further focus.

It is easily seen from the above that the excentricity is least when Q is in the line SP and greatest when Q is in the line SP produced. Hence the maximum and minimum values of the excentricity are $M_1 + M_2$ and $M_1 \sim M_2$ respectively. (Art. 74.)

79. *To integrate the equations for the inclination and longitude of the node.*

We have (Art. 61) for the planet m

$$\frac{di}{dt} = \frac{m'na^2a'}{4\mu} D_1 \tan i'' \sin (\Omega - \Omega'),$$

$$\tan i \cdot \frac{d\Omega}{dt} = - \frac{m'na^2a'}{4\mu} D_1 \{\tan i - \tan i'' \cos (\Omega - \Omega')\};$$

with similar equations for the planet m'.

To integrate these, assume

$$p = \tan i \sin \Omega, \quad q = \tan i \cos \Omega,$$
$$p' = \tan i'' \sin \Omega', \quad q' = \tan i'' \cos \Omega';$$

therefore $\dfrac{dp}{dt} = \tan i \cos \Omega \, \dfrac{d\Omega}{dt} + \sin \Omega \, (1 + \tan^2 i) \dfrac{di}{dt}.$

Substituting the expressions for $\dfrac{d\Omega}{dt}$ and $\dfrac{di}{dt}$, and writing α for $\dfrac{m'na^2a'}{4\mu}$, since $\tan^2 i \dfrac{di}{dt}$ being of the third order may be omitted, we have

$$\frac{dp}{dt} = \alpha D_1 \, (\tan i'' \cos \Omega' - \tan i \cos \Omega)$$

$$= \alpha D_1 \, (q' - q).$$

Similarly, $\qquad \dfrac{dq}{dt} = \alpha D_1 \, (p - p').$

Also for the planet m', writing α' for $\dfrac{mn'a'^2a}{4\mu}$,

$$\frac{dp'}{dt} = \alpha' D_1 (q - q'),$$

$$\frac{dq'}{dt} = \alpha' D_1 \, (p' - p).$$

The forms of these equations suggest the following particular integrals:

$$p = N \sin (ht + \delta), \qquad q = N \cos (ht + \delta),$$
$$p' = N' \sin (ht + \delta), \qquad q' = N' \cos (ht + \delta).$$

Substituting these in the differential equations, we obtain from either of the first two,

$$hN = aD_1 (N' - N),$$

and from either of the last two,

$$hN' = a'D_1 (N - N'),$$

eliminating the ratio $N : N'$,

$$(h + aD_1) (h + a'D_1) = aa'D_1^2,$$

or

$$h^2 + (a + a') D_1 h = 0;$$

therefore

$$h = -(a + a') D_1, \text{ or } h = 0.$$

Denote the former by h_1, and let δ_1, δ_2, N_1, N_2, N_1', N_2', be the values of δ, N, N' corresponding to $h = h_1$ and $h = 0$. Then $N_2' = N_2$, and the complete solution of the differential equations will be

$$p = N_1 \sin (h_1 t + \delta_1) + N_2 \sin \delta_2,$$
$$q = N_1 \cos (h_1 t + \delta_1) + N_2 \cos \delta_2,$$
$$p' = N_1' \sin (h_1 t + \delta_1) + N_2 \sin \delta_2,$$
$$q' = N_1' \cos (h_1 t + \delta_1) + N_2 \cos \delta_2.$$

Of the constants in these equations, four are arbitrary, and must be determined from the known values of i and Ω at some given epoch.

We have then

$$\tan^2 i = p^2 + q^2 = N_1^2 + N_2^2 + 2N_1 N_2 \cos (h_1 t + \delta_1 - \delta_2),$$
$$\tan \Omega = \frac{p}{q} = \frac{N_1 \sin (h_1 t + \delta_1) + N_2 \sin \delta_2}{N_1 \cos (h_1 t + \delta_1) + N_2 \cos \delta_2},$$

with similar equations for i' and Ω'.

C. P. T.

6

Had we considered a system of several planets, we should have obtained a result precisely similar to that of Art. 73.

80. From the form of the expression for tan i, the stability of the inclinations may be inferred. For it may be shewn, as in Art. 74, that tan i can never exceed

$$N_1 + N_2 + N_3 + \dots,$$

these quantities being taken with the same sign. Since then N_1, N_2, N_3 &c. are found to be very small, it follows that the inclinations must always remain exceedingly small.

In the case of two mutually disturbing planets, we learn from the expression in Art. 79, that tan i fluctuates between the limits $N_1 + N_2$ and $N_1 \sim N_2$. The periods of these changes are the same for the two planets, being $\dfrac{2\pi}{\pm h_1}$; and as appears from the equation of Art. 66, the maximum of each inclination will take place at the time of the minimum of the other.

In the case of Jupiter and Saturn, the period is 50673 years; the maximum and minimum inclinations of Jupiter's orbit to the ecliptic are $2^\circ 2' 30''$ and $1^\circ 17' 10''$, those of Saturn's orbit $2^\circ 32' 40''$ and $0^\circ 47'$.

81. We now proceed to examine the expression which has been obtained in Art. 79 for the longitude of the node. We have

$$\tan \Omega = \frac{N_1 \sin (h_1 t + \delta_1) + N_2 \sin \delta_2}{N_1 \cos (h_1 t + \delta_1) + N_2 \cos \delta_2}.$$

The maxima and minima values of Ω, if such exist, will be found by equating $\dfrac{d\Omega}{dt}$ to zero. Thus

$$\cos (h_1 t + \delta_1 - \delta_2) = - \frac{N_1}{N_2}.$$

If this (disregarding sign) be not greater than unity, the node will oscillate, the period of a complete oscillation being the same as that of the inclinations, viz. $\dfrac{2\pi}{\pm h_1}$. But if it be greater than unity, there cannot be any stationary positions, and the node will move continually in one direction.

It may be shown, as in Art. 76, that the motion of the node will be fastest or slowest whenever the inclination is either a maximum or a minimum.

82. *When the line of nodes oscillates, to find the extent and periods of its oscillations.*

It may be shewn as in Art. 77, that if ψ be written for $h_1 t + \delta_1 - \delta_2$, and τ denote the least positive angle whose cosine is $\dfrac{N_1}{N_2}$,

$$\tan (\Omega - \delta_2) = \frac{N_1 \sin \psi}{N_2 + N_1 \cos \psi}$$

$$= \frac{\cos \tau \sin \psi}{1 + \cos \tau \cos \psi},$$

and

$$\tan^2 i \; \frac{d\Omega}{dt} = h_1 N_1 N_2 (\cos \psi + \cos \tau).$$

Different cases will occur according to the signs of N_1, N_2 and h_1. Suppose N_1, N_2 of like sign, and h_1 positive: then ψ increases as t increases, and the line of nodes regredes so long as ψ is between $(2n-1)\pi - \tau$ and $(2n-1)\pi + \tau$, and progredes so long as ψ is between $(2n-1)\pi + \tau$ and $(2n+1)\pi - \tau$.

Let Ω', Ω'' be the values of Ω corresponding to the values $(2n-1)\pi - \tau$ and $(2n-1)\pi + \tau$ of ψ; then

$$\tan (\Omega' - \delta_2) = \cot \tau,$$

$$\tan (\Omega'' - \delta_2) = -\cot \tau;$$

therefore
$$\Omega' - \delta_2 = m\pi + \frac{\pi}{2} - \tau,$$

$$\Omega'' - \delta_2 = (m-1)\pi + \frac{\pi}{2} + \tau;$$

therefore
$$\Omega' - \Omega'' = \pi - 2\tau,$$

which is the angle through which the line of nodes regredes. Also the period of this regression may be shewn as in Art. 77 to be $\frac{2\tau}{h_1}$. Similarly, the angle through which the line of nodes progredes may be shewn to be $\pi - 2\tau$, and the period of the progression $\frac{2(\pi - \tau)}{h_1}$.

The period of a complete oscillation will be the sum of the periods of the regression and progression, that is $\frac{2\pi}{h_1}$, which agrees with the preceding Article.

The remaining cases corresponding to different arrangements of the signs of N_1, N_2 and h_1 may be treated in like manner.

The mean value of Ω is $m\pi + \delta_2$, whatever be the signs of N_1, N_2 and h_1, and the mean value coincides with the true whenever $\sin\psi = 0$. Since, then, ψ is the same both for the disturbed and disturbing planet, the nodes of both orbits will arrive simultaneously at their mean positions.

In the case of Jupiter and Saturn, N_2 is for each planet numerically less than N_1, so that the node oscillates; the extent of oscillation being $13° 9' 40''$ in Jupiter's orbit, and $31° 56' 20''$ in that of Saturn on either side of their mean position, the ecliptic being taken for the plane of reference, and supposed immoveable.

83. *To shew that the inclination of the orbits of two mutually disturbing planets to each other is approximately constant.*

If γ denote this inclination, we have by Spherical Trigonometry,

$$\cos \gamma = \cos i \cos i' + \sin i \sin i' \cos (\Omega - \Omega')$$
$$= \cos i \cos i' \{1 + \tan i \tan i' \cos (\Omega - \Omega')\}$$
$$= (1 + \tan^2 i)^{-\frac{1}{2}} (1 + \tan^2 i')^{-\frac{1}{2}} \{1 + \tan i \tan i' \cos (\Omega - \Omega')\}$$
$$= 1 - \frac{1}{2} \{\tan^2 i + \tan^2 i' - 2 \tan i \tan i' \cos (\Omega - \Omega')\},$$

if we neglect small quantities of orders higher than the second.

Now $\tan^2 i + \tan^2 i' - 2 \tan i \tan i' \cos (\Omega - \Omega')$
$$= p^2 + q^2 + p'^2 + q'^2 - 2 (pp' + qq')$$
$$= (p - p')^2 + (q - q')^2$$
$$= (N_1 - N_1')^2;$$

therefore $1 - \cos \gamma = \frac{1}{2} (N_1 - N_1')^2,$

or $\sin \frac{\gamma}{2} = \frac{1}{2} (N_1 \sim N_1');$

whence it follows that γ is constant.

84. The equations which give the secular variations of the node and inclination may be explained geometrically as follows*.

The equations to be interpreted are
$$p = N_1 \sin (h_1 t + \delta_1) + N_2 \sin \delta_2,$$
$$q = N_1 \cos (h_1 t + \delta_1) + N_2 \cos \delta_2,$$
where $p = \tan i \sin \Omega, \quad q = \tan i \cos \Omega.$

* This elegant geometrical explanation is due to Mr H. M. Taylor, M.A., Fellow and Tutor of Trinity College, Cambridge. *Oxford, Cambridge and Dublin Messenger of Mathematics*, Vol. III. p. 189.

Let NA, NB be the intersections of the planes of reference and of the orbit respectively with a sphere of radius
unity, the centre of which coincides with that of the Sun:
$ABZP$ the great circle of which the pole is N: Z and P the
poles of the great circles NA, NB: L the point from which
longitude is measured.

Then $LN = \Omega$, and $PZ = BA = i$.

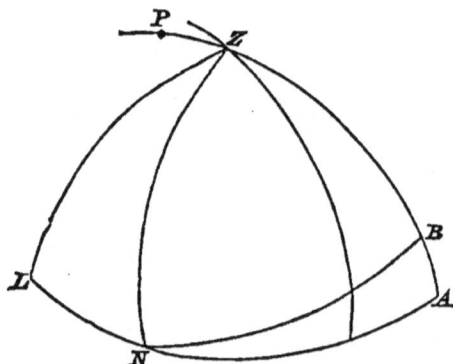

Now project PZ and the other great circles by radii
drawn from the centre of the sphere on the tangent plane at
Z: then $\tan i \sin \Omega$ and $\tan i \cos \Omega$ are the Cartesian co-
ordinates of the projection of P referred to the projection
of LZ as axis of y, and a line at right angles to it as axis
of x.

Suppose P' the projection of P, then if the co-ordinates
of P' be x and y, we have

$$x = N_1 \cos (h_1 t + \delta_1) + N_2 \cos \delta_2,$$
$$y = N_1 \sin (h_1 t + \delta_1) + N_2 \sin \delta_2.$$

These equations shew us that P' always lies on a circle of
which the centre is at the fixed point $(N_2 \cos \delta_2,\ N_2 \sin \delta_2)$,
and the radius is N_1: also that P' describes this circle with a
uniform angular velocity h_1.

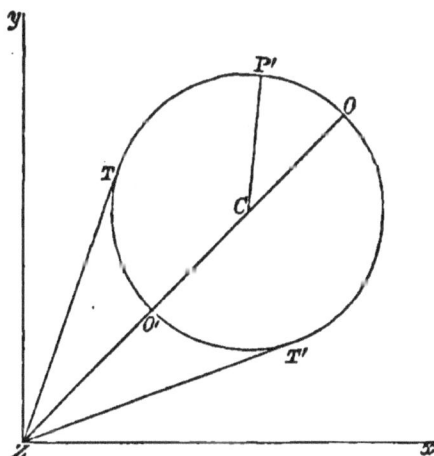

We may hence arrive geometrically at the results of Arts. 81 and 82. If C be the centre of the circle, we have

$$CP' = N_1, \quad ZC = N_2,$$

angle $CZx = \delta_2$, angle $P'CO = h_1 t + \delta_1 - \delta_2$;

and if $P'Z$ be joined,

$$P'Z = i, \text{ angle } P'Zx = \Omega.$$

Suppose N_1, N_2 both positive; and first, let N_1 be less than N_2.

Now the only time when the node will be stationary will be when P' is moving directly towards or directly from Z, that is at T, T' the points of contact of tangents to the circle drawn from Z. As P' moves from T' to T, Ω increases, or the node progredes, and as P' moves from T to T', Ω decreases, or the node regredes.

If, then, τ denote the angle $TCZ\left(\cos^{-1}\frac{N_1}{N_2}\right)$, the period of the progression will be the time P' takes to move from T' to T, that is $\frac{2(\pi - \tau)}{h_1}$; the period of the regression will be the time P' takes to move from T to T', that is $\frac{2\tau}{h_1}$; and the node is stationary whenever $h_1 t + \delta_1 - \delta_2 = (2n + 1)\,\pi \pm \tau.$ Also the angle through which the node oscillates is TZT' which $= \pi - 2\tau.$

Secondly, suppose $N_1 = N_2.$ Then Z will lie on the circumference of the circle which P' traces out.   In this case T and T' coincide, and the node after becoming stationary begins to move in the same direction as before.

Thirdly, suppose N_1 greater than $N_2.$ Then Z lies within the circle, and no tangents can be drawn from it to the circle: from this we see that the node is never stationary.

It is easily seen that in all three cases the maximum and minimum values of the inclination are ZO and ZO', that is $N_1 + N_2$ and $N_1 \sim N_2$ respectively.   And in all cases the node moves fastest or slowest when P' coincides with O or O', that is, whenever the inclination is a maximum or minimum.

The above geometrical construction also affords a proof of the theorem of Art. 83.

Since in the case of two mutually disturbing planets the quantities h_1, δ_1, N_2, δ_2 are the same for both, it follows that the point P'' for the second planet traces out a circle concentric with that traced out by P', and with the same uniform angular velocity h_1; also that the two points P', P'' always lie in a common radius vector through C.   Now the angle subtended by $P'P''$ at the centre of the sphere is the inclination of the two orbits; this inclination is therefore very nearly constant, as $P'P''$ is small and constant and very near Z.

85.   *To integrate the equation for the longitude of the epoch.*

We have (Art. 61)

$$\frac{d\epsilon}{dt} = A + A_1\,(e^2 - \tan^2 i) + A_2\,(e'^2 - \tan^2 i')$$
$$+ A_3 ee' \cos(\varpi - \varpi') + A_4 \tan i \tan i' \cos(\Omega - \Omega').$$

Now from the formulæ of Art. 72, we obtain

$$e^2 = M_1^2 + M_2^2 + 2M_1M_2 \cos\{(g_1 - g_2) t + \gamma_1 - \gamma_2\},$$

$$e'^2 = M_1'^2 + M_2'^2 + 2M_1'M_2' \cos\{(g_1 - g_2) t + \gamma_1 - \gamma_2\},$$

$$ee' \cos(\varpi - \varpi') = M_1M_1' + M_2M_2'$$

$$+ (M_1M_2' + M_2M_1') \cos\{(g_1 - g_2) t + \gamma_1 - \gamma_2\}.$$

In like manner, from the formulæ of Art. 79,

$$\tan^2 i = N_1^2 + N_2^2 + 2N_1N_2 \cos\{h_1 t + \delta_1 - \delta_2\},$$

$$\tan^2 i' = N_1'^2 + N_2'^2 + 2N_1'N_2 \cos\{h_1 t + \delta_1 - \delta_2\},$$

$$\tan i \tan i'' \cos(\Omega - \Omega') = N_1N_1' + N_2^2$$

$$+ N_2 (N_1 + N_1') \cos\{h_1 t + \delta_1 - \delta^2\}.$$

If these values be substituted in the expression for $\dfrac{d\epsilon}{dt}$, it takes the form

$$\frac{d\epsilon}{dt} = Bn + B_1 \cos\{(g_1 - g_2) t + \gamma_1 - \gamma_2\} + B_2 \cos\{h_1 t + \delta_1 - \delta_2\},$$

where Bn, B_1, and B_2 denote certain constants. Integrating, we have

$$\epsilon = \epsilon_0 + Bnt + \frac{B_1}{g_1 - g_2} \sin\{(g_1 - g_2) t + \gamma_1 - \gamma_2\}$$

$$+ \frac{B_2}{h_1} \sin\{h_1 t - \delta_1 - \delta_2\}.$$

We may omit the term Bnt, if we consider it as furnishing a correction on the mean motion n, which thus becomes $(1 + B) n$. With this understanding

$$\epsilon = \epsilon_0 + \frac{B_1}{g_1 - g_2} \sin\{(g_1 - g_2) t + \gamma_1 - \gamma_2\} + \frac{B_2}{h_1} \sin\{h_1 t + \delta_1 - \delta_2\}.$$

If this expression be developed, we may again omit the term involving the first power of t, and consider it as affording a further correction to the mean motion*. Thus we shall obtain a series of the form

$$\delta\epsilon = B_3 t^2 + B_4 t^3 + \dots$$

86. In the Theory of the Planets this inequality is insensible, but in that of the Moon it amounts to upwards of 10 seconds in a century, forming what is termed the *secular acceleration of the Moon's mean motion*. Thus it appears that this inequality does not, as its name would seem to imply, contradict the general theorem of the invariability of the mean motions, since it is due to a variation, not of the mean motion, (as we have employed the term in the preceding pages,) but of the epoch. If, however, as in the Lunar Theory, the epoch be omitted, any variation in the mean longitude will of necessity be thrown upon the mean motion; only in this case, n will not be given by the equation $n^2 a^3 = \mu$.

87. We have hitherto supposed the planetary motions to be referred to a fixed plane, but have left the particular plane undetermined. In practice it is usual to take the position of the ecliptic at some given epoch, as for instance the year 1800; but since it is to the true ecliptic that astronomers refer the celestial motions, we will now obtain formulæ for determining relatively to the plane of the Earth's orbit, the position of that of any other planet.

Let then m, m' denote the masses of the Earth and the planet considered, and suppose the orbits of m and m' but little inclined to each other and to the fixed plane of reference. Let λ, λ' denote the latitudes of points in these orbits corresponding to the same longitude θ_1; then (see fig. to Art. 13)

$$\tan \lambda = \tan i \sin (\theta_1 - \Omega), \quad \tan \lambda' = \tan i'' \sin (\theta_1 - \Omega').$$

* The advantage of thus disposing of these terms arises from the fact that the mean motion, as determined by observation, is the complete coefficient of t in the expression of the mean longitude.

Now since i and i' are very small, we may replace $\tan\lambda$, $\tan\lambda'$ by λ, λ' respectively: thus

$$\lambda' - \lambda = \tan i'' \sin(\theta_1 - \Omega') - \tan i \sin(\theta_1 - \Omega)$$
$$= (\tan i' \cos \Omega' - \tan i \cos \Omega) \sin \theta_1$$
$$- (\tan i'' \sin \Omega' - \tan i \sin \Omega) \cos \theta_1,$$

or, with the notation of Art. 70,

$$\lambda' - \lambda = (q' - q) \sin \theta_1 - (p' - p) \cos \theta_1 \ldots\ldots\ldots (1).$$

Now let γ denote the inclination, ν the longitude of the node of the orbit of m' relatively to that of m; then approximately

$$\lambda' - \lambda = \tan \gamma \sin(\theta_1 - \nu)$$
$$= \tan \gamma \cos \nu \sin \theta_1 - \tan \gamma \sin \nu \cos \theta_1, \ldots\ldots\ldots(2).$$

Hence equating coefficients of $\sin\theta_1$ and $\cos\theta_1$ in equations (1) and (2),

$$\tan \gamma \cos \nu = q' - q, \quad \tan \gamma \sin \nu = p' - p;$$

whence $\qquad \tan^2 \gamma = (p' - p)^2 + (q' - q)^2,$

and $\qquad\qquad\qquad \tan \nu = \dfrac{p' - p}{q' - q}.$

These expressions determine the position of the orbit of m' relatively to that of m, when the values of p, p', q, and q' are known. Differentiating them, and neglecting small quantities of orders higher than the second, we obtain

$$\frac{d\gamma}{dt} = \left(\frac{dp'}{dt} - \frac{dp}{dt}\right) \sin \nu + \left(\frac{dq'}{dt} - \frac{dq}{dt}\right) \cos \nu,$$

$$\frac{d\nu}{dt} = \left(\frac{dp'}{dt} - \frac{dp}{dt}\right) \frac{\cos \nu}{\tan \gamma} - \left(\frac{dq'}{dt} - \frac{dq}{dt}\right) \frac{\sin \nu}{\tan \gamma}.$$

If the values of $\dfrac{dp}{dt}$, $\dfrac{dq}{dt}$, &c. be substituted, these equations give the variations of γ and ν.

88.   It has been proposed to employ the invariable plane of the Solar System as a plane of reference ; and since it remains absolutely fixed, it would afford a means of comparing together observations separated by long intervals. But the want of a convenient fixed line upon it, from which to measure longitudes*, stands in the way of its practical application ; so that it must be looked upon, at all events for the present, rather as theoretically interesting than as available for astronomical purposes.

* See Pontécoulant's *Système du Monde*, tome I. p. 469 (2nd edit.).

Note.   Gauss in a memoir (*Determinatio Attractionis, &c.*) contributed to the Transactions of the Royal Society of Science of Gottingen, 1818, has stated without proof the following theorem.   The secular variations which the elements of a planetary orbit experience from the perturbation produced by another planet are independent of the position of this planet in its orbit, and would be the same whether the disturbing planet moved in its elliptic orbit according to Kepler's Laws, or whether its mass be conceived to be equably distributed throughout the orbit in such manner that equal parts of the mass are now assigned to such parts of the orbit as were described in equal intervals of time, provided that the times of revolution of the disturbed and disturbing planets are not commensurable.   Gauss has also shewn how to reduce the expressions for the attraction of such a mass-orbit at any assigned mass-point not on the orbit to calculable forms of elliptic integrals.   [A. F.]

CHAPTER VI.

PERIODIC VARIATIONS OF THE ELEMENTS OF THE ORBIT.

89. WE come now to consider the variations produced by the periodical terms of R. These are called *Periodical Variations*, as opposed to the Secular Variations produced by the non-periodic terms. We have seen, indeed, that the latter are for the most part periodical in form, but in the Planetary Theory, the term Periodical Variations is restricted to those we are about to consider in the present Chapter.

90. We have seen (Art. 46) that the general type of a periodical term is $P \cos\{(pn \pm qn')t + Q\}$, where P is a function of a, e, i; and Q is a function of ϖ, ϵ, Ω. Now such a term will produce a similar term in $\dfrac{dR}{da}$, $\dfrac{dR}{de}$, and $\dfrac{dR}{di}$; but a term of the form $P \sin\{(pn \pm qn')t + Q\}$ in $\dfrac{dR}{d\varpi}$, $\dfrac{dR}{d\epsilon}$, and $\dfrac{dR}{d\Omega}$. If, then, these be substituted in the equations of Art. 39, they will take the forms

$$\frac{da}{dt} = P_1 \sin\{(pn \pm qn')t + Q\},$$

$$\frac{de}{dt} = P_2 \sin\{(pn \pm qn')t + Q\},$$

$$\frac{d\varpi}{dt} = P_3 \cos \{(pn \pm qn')\, t + Q\},$$

$$\frac{d\epsilon}{dt} = P_4 \cos \{(pn \pm qn')\, t + Q\},$$

$$\frac{d\Omega}{dt} = P_5 \cos \{(pn \pm qn')\, t + Q\},$$

$$\frac{di}{dt} = P_6 \sin \{(pn \pm qn')\, t + Q\},$$

$$\frac{d^2\zeta}{dt^2} = P_7 \sin \{(pn \pm qn')\, t + Q\},$$

where P_1, P_2, &c. are functions of the elements of the disturbed and disturbing planets, and involve the first power of the disturbing mass.

91. In integrating these equations, we may in general consider the elements which enter in the right-hand members as constant and equal to their values at the epoch from which the time is reckoned[*].

Let then a, e, ϖ, &c. denote the values of the elements at epoch, δa, δe, $\delta \varpi$, &c. their periodical variations after an interval t: then integrating the above equations, and omitting the constant terms, we have

$$\delta a = - \frac{P_1}{pn \pm qn'} \cos \{(pn \pm qn')\, t + Q\},$$

$$\delta e = - \frac{P_2}{pn \pm qn'} \cos \{(pn \pm qn')\, t + Q\},$$

$$\delta \varpi = \frac{P_3}{pn \pm qn'} \sin \{(pn \pm qn')\, t + Q\},$$

* This is equivalent to neglecting the square of the disturbing force: see Art. 95.

$$\delta\epsilon = \frac{P_4}{pn \pm qn'}, \sin\{(pn \pm qn')\, t + Q\},$$

$$\delta\Omega = \frac{P_5}{pn \pm qn'}, \sin\{(pn \pm qn')\, t + Q\},$$

$$\delta i = -\frac{P_6}{pn \pm qn'}, \cos\{(pn \pm qn')\, t + Q\},$$

$$\delta\zeta = -\frac{P_7}{(pn \pm qn')^2}, \sin\{(pn \pm qn)\, t + Q\}.$$

Hence it appears that the variations produced by the periodical terms of R are all periodical in form.

92. It will be seen that all the expressions of the last Article involve the divisor $pn \pm qn'$, while $\delta\zeta$ involves the divisor $(pn \pm qn')^2$. If then it should happen that either $pn + qn'$ or $pn - qn'$ is very small, a term in R containing $(pn \pm qn')\, t$ in its argument, though of a high order, may have a sensible effect on the elements of the orbit. Now since p and q are either positive integers or zero, $pn + qn'$ cannot be small unless n and n' are small, a case which does not occur with any of the planets: but we have instances in which $pn - qn'$ is small*.

Since the period of such inequalities is very great $\Big($being $\dfrac{2\pi}{pn - qn'}\Big)$, they are called *inequalities of long period*, or *long inequalities*.

93. *To select such terms in* R *as will produce the principal inequalities of long period.*

---

* If in any case the mean motions of two planets were exactly commensurable and in the ratio of p to q, the corresponding term of R, as we have already remarked (Art. 48), would cease to be periodical and would form a part of F, but no instance of this occurs among the planets.

We have seen that the dimension of the principal part of the coefficient of a term containing $(pn - qn')t$ in its argument is $p - q$ (Art. 50); hence if we can find two integers p and q nearly in the ratio of n' to n, and having a small difference, the corresponding term of R will produce an important long inequality in the elements of each planet.

In the case of Jupiter and Saturn $n : n' :: 5 : 2$ nearly, and $5 - 2 = 3$; hence there is a long inequality arising from a term in R of the form $P \cos \{(2n - 5n') t + Q\}$, the principal part of P being of the third order. This inequality is interesting in an historical point of view, having long baffled the labours of mathematicians and appeared inexplicable on the hypothesis of gravitation. It was at last successfully explained by Laplace.

For the Earth and Venus, $n : n' :: 8 : 13$ nearly, so that there is a long inequality arising from a term in R of the fifth order. The discovery of this inequality is due to the Astronomer Royal.

Finally, in the case of Neptune and Uranus, $n : n' :: 1 : 2$ nearly, hence there is a long inequality arising from a term in R which is of the first order.

94. Between corresponding terms of the long inequalities in the mean motions of two planets, arising from the near commensurability of n and n', there is a simple approximate relation.

Let m, m' be the masses of the two planets, R, R' their disturbing functions: then by Art. 8, considering only the mutual action of m and m', we have

$$R = \frac{m'}{\rho'} - \frac{m'}{r'^3} (xx' + yy' + zz'),$$

$$R' = \frac{m}{\rho'} - \frac{m}{r^3}(xx' + yy' + zz').$$

We shall distinguish the first and second terms of R and R' as the *symmetrical* and *unsymmetrical* parts respectively, since the co-ordinates of m and m' are involved symmetrically in the former but not in the latter.

Since, then, the symmetrical parts of R and R' differ only in having m and m' interchanged, if

$$m'M \cos\{(pn - qn')t + Q\}$$

be any term in the symmetrical part of R, that of R' will contain the term

$$mM \cos\{(pn - qn')t + Q\}.$$

Confining our attention to these terms, we have (Art. 39)

$$\frac{d^2\zeta}{dt^2} = -\frac{3n^2 a}{\mu}\frac{dR}{d\epsilon} = -\frac{3na}{\mu}\frac{d(R)}{dt}$$

$$= \frac{3n^2 ap}{\mu} m'M \sin\{(pn - qn')t + Q\},$$

therefore $\delta\zeta = -\dfrac{3n^2 ap}{\mu}\dfrac{m'M}{(pn - qn')^2}\sin\{(pn - qn')t + Q\}.$

Similarly, $\delta\zeta' = \dfrac{3n'^2 a' q}{\mu'}\dfrac{mM}{(pn - qn')^2}\sin\{(pn - qn')t + Q\}.$

Hence $\dfrac{\delta\zeta}{\delta\zeta'} = -\dfrac{m'\mu'n^2 ap}{m\mu n'^2 a' q} = -\dfrac{m'\mu'na}{m\mu n'a'}$

approximately, since qn' is nearly equal to pn; therefore

$$\frac{\delta\zeta}{\delta\zeta'} = -\frac{m'\sqrt{(\mu'a')}}{m\sqrt{(\mu a)}},$$

or, since μ' differs from μ by a quantity of the order of the disturbing force, the square of which we are neglecting, we have to the first order,

$$\frac{\delta \zeta}{\delta \zeta'} = - \frac{m' \sqrt{a'}}{m \sqrt{a}},$$

the required relation.

The same relation is also approximately true in the case of terms arising from the unsymmetrical parts of R and $R'*$. For denoting these by R_1 and R_1' respectively, we have

$$R_1 = - m' \left(x \frac{x'}{r'^3} + y \frac{y'}{r'^3} + z \frac{z'}{r'^3} \right),$$

$$R_1' = - m \left(x' \frac{x}{r^3} + y' \frac{y}{r^3} + z' \frac{z}{r^3} \right).$$

Now the equations of motion of the planet m' referred to rectangular axes are

$$\frac{d^2 x'}{dt^2} + \frac{\mu' x'}{r'^3} = \frac{dR'}{dx'}, \text{ \&c.;}$$

and since, by the principles of the method of the variation of elements, the *form* of solution of these equations is the same as it would be if R' were zero, it follows that, if the differential coefficients be taken *as if the elements were constant*, we shall have

$$\frac{x'}{r'^3} = - \frac{1}{\mu'} \frac{d^2 x'}{dt^2}, \quad \frac{y'}{r'^3} = - \frac{1}{\mu'} \frac{d^2 y'}{dt^2}, \quad \frac{z'}{r'^3} = - \frac{1}{\mu'} \frac{d^2 z'}{dt^2};$$

and therefore, with this understanding, that

$$R_1 = \frac{m'}{\mu'} \left(x \frac{d^2 x'}{dt^2} + y \frac{d^2 y'}{dt^2} + z \frac{d^2 z'}{dt^2} \right).$$

* For the demonstration of this we are mainly indebted to *The Theory of the Long Inequality of Uranus and Neptune:* an essay which obtained the Adams Prize for the year 1850. By R. Pierson, M.A.

Similarly, $\quad R_1' = \dfrac{m}{\mu}\left(x'\dfrac{d^2x}{dt^2} + y'\dfrac{d^2y}{dt^2} + z'\dfrac{d^2z}{dt^2}\right)$.

Now any term in R_1 containing $(pn - qn')t$ in its argument can arise only from the combination of terms in x, y, and z, containing pnt in their arguments with terms in $\dfrac{d^2x'}{dt^2}$, $\dfrac{d^2y'}{dt^2}$, and $\dfrac{d^2z'}{dt^2}$ containing $qn't$. Suppose, then, x and x' when developed in terms of t and the elements to contain respectively the terms

$$L\cos(pnt + l), \quad L'\cos(qn't + l').$$

Hence the product $x\dfrac{d^2x'}{dt^2}$ will contain the term

$$-\frac{1}{2}LL'q^2n'^2\cos\{(pn - qn')t + l - l'\},$$

and the product $x'\dfrac{d^2x}{dt^2}$ the term

$$-\frac{1}{2}LL'p^2n^2\cos\{(pn - qn')t + l - l'\}:$$

the coefficients are in the ratio $q^2n'^2$ to p^2n^2. Similarly, the coefficients of the same cosine in $y\dfrac{d^2y'}{dt^2}$ and $z\dfrac{d^2z'}{dt^2}$ are to those in $y'\dfrac{d^2y}{dt^2}$ and $z'\dfrac{d^2z}{dt^2}$ in the same ratio.

Hence if $\quad \dfrac{m'}{\mu'}Mq^2n'^2\cos\{(pn - qn')t + Q\}$

be any term in R_1, then R_1' will contain the term

$$\frac{m}{\mu}Mp^2n^2\cos\{(pn - qn')t + Q\}.$$

Confining our attention to these terms, we have

$$\delta\zeta = -\frac{3n^2ap}{\mu\mu'}\frac{m'Mq^2n'^2}{(pn-qn')^2}\sin\{(pn-qn')t+Q\},$$

$$\delta\zeta' = \frac{3n'^2a'q}{\mu\mu'}\frac{mMp^2n^2}{(pn-qn')^2}\sin\{(pn-qn')t+Q\}.$$

Hence

$$\frac{\delta\zeta}{\delta\zeta'} = -\frac{m'aq}{ma'p} = -\frac{m'na}{mn'a'}\text{ nearly,}$$

$$= -\frac{m'\sqrt{a'}}{m\sqrt{a}},$$

the square of the disturbing force being neglected.

By means of this relation, when one of the long inequalities is known, the other may be calculated : it may also be used as a formula of verification.

95.   We have remarked that in integrating the equations of Art. 90, we may in general consider the elements which enter in the right-hand members as constant and equal to their values at the epoch from which the time is measured. In the case, however, of inequalities whose periods are very long, the secular variations of the elements in the interval produce a sensible effect.   In order to take account of these, we may integrate our equations by parts, considering the elements variable ; and then substitute their values as calculated by the method of Art. 62.   For example, consider the equation

$$\frac{d^2\zeta}{dt^2} = P\sin\{(pn-qn')t+Q\}$$

$$= P\sin\lambda,\text{ suppose.}$$

Integrating by parts, and remembering that n is constant with regard to secular variations, we have

$$\frac{d\zeta}{dt} = -\frac{P}{pn-qn'}\cos\lambda + \frac{1}{(pn-qn')^2}\frac{dP}{dt}\sin\lambda$$

$$+ \frac{1}{(pn-qn')^3}\frac{d^2P}{dt^2}\cos\lambda - \ldots;$$

therefore $\delta\zeta = -\dfrac{P}{(pn-qn')^2}\sin\lambda - \dfrac{1}{(pn-qn')^3}\dfrac{dP}{dt}\cos\lambda$

$$+ \frac{1}{(pn-qn')^4}\frac{d^2P}{dt^2}\sin\lambda + \ldots$$

$$- \frac{1}{(pn-qn')^3}\frac{dP}{dt}\cos\lambda + \frac{1\cdot}{(pn-qn')^4}\frac{d^2P}{dt^2}\sin\lambda + \ldots$$

$$+ \frac{1}{(pn-qn')^4}\frac{d^2P}{dt^2}\sin\lambda + \ldots$$

$$= \left\{-\frac{P}{(pn-qn')^2} + \frac{3}{(pn-qn')^4}\frac{d^2P}{dt^2} - \ldots\right\}\sin\lambda$$

$$+ \left\{-\frac{2}{(pn-qn')^3}\frac{dP}{dt} - \ldots\right\}\cos\lambda.$$

In this equation P, $\dfrac{dP}{dt}$, $\dfrac{d^2P}{dt^2}$, &c. are functions of the elements; their values may be calculated by the formulæ of Art. 62. It may be noticed that P is of the first order, $\dfrac{dP}{dt}$ of the second, and $\dfrac{d^2P}{dt^2}$ of the third of the disturbing force: for $\dfrac{dP}{dt}$, being found from P by differentiation, will involve the differential coefficients of the elements, which are themselves of the first order; and similarly for $\dfrac{d^2P}{dt^2}$.

96. Having now completed our account of the. methods

of treating the secular and periodic variations of the elements
of the orbit, we will say a few words on the distinction between them.   In the first place we may observe that the
periodic variations involve the mean longitude of the disturbed and disturbing planets, and therefore depend chiefly
upon the configuration of the planetary system.   On the contrary the secular variations depend solely upon the values of
the elements.   The latter class of variations take place with
extreme slowness, so that if these only existed, a considerable
time must elapse before the deviation of the planet from
elliptic motion became appreciable.   On the other hand, the
periodic variations (such at least as are rapidly periodic)
"are in their nature transient and temporary: they disappear
in short periods, and leave no trace.   The planet is tempo-
rarily drawn from its orbit (its slowly varying orbit), but
forthwith returns to it, to deviate presently as much the other
way, while the varied orbit accommodates and adjusts itself
to the average of these excursions on either side of it; and
thus continues to present, for a succession of indefinite ages,
a kind of medium picture of all that the planet has been
doing in their lapse, in which the expression and character
is preserved; but the individual features are merged and
lost*."   On this account it is convenient to suppose the
planet to move in an ellipse, the elements of which are corrected for secular variations only, and to take account of the
periodic variations by applying small corrections to the radius
vector and longitude as calculated from the elliptic formulæ.

97.   We will accordingly shew how by means of the
periodic variations of the elements, the corresponding inequalities in the radius vector and longitude may be calculated.   If we take for our plane of reference the position of

* Herschel's *Outlines of Astronomy*, 10th edit. Art. 656.

the plane of the orbit of the disturbed planet at the epoch from which the time is reckoned, the inclination will be of the order of the disturbing force, and therefore, if we neglect the square of the latter, we may also neglect the square of the former.

98. *To calculate the periodic variations in radius vector.*

Let δa, δe, $\delta \varpi$, &c. denote the periodic variations in a, e, ϖ, &c., and let δr be the corresponding variation in r; then

$$\delta r = \frac{dr}{da} \delta a + \frac{dr}{de} \delta \epsilon + \frac{dr}{d\varpi} \delta \varpi + \frac{dr}{d\zeta} \delta \zeta + \frac{dr}{d\epsilon} \delta \epsilon,$$

in which the square of the disturbing force is neglected, since this would be introduced by the squares and products of δa, δe, &c. The values of δa, δe, &c. have been found in Art. 91, those of $\frac{dr}{da}$, $\frac{dr}{de}$, &c. may be obtained from the equation (Art. 40)

$$r = a \left\{ 1 + \frac{1}{2} e^2 - e \cos (\zeta + \epsilon - \varpi) - \frac{1}{2} e^2 \cos 2 (\zeta + \epsilon - \varpi) - \ldots \right\}.$$

99. *To calculate the periodic variations in longitude.*

These might be found in the same manner as the variations in radius vector, but they may also be deduced from them: we proceed to obtain a formula for this purpose.

We have $\qquad \dfrac{d\theta_0}{dt} = \dfrac{h}{r^2}$, (Art. 22),

and $\qquad \theta - \theta_0 = \Omega - \Omega_0;$

therefore $\qquad \dfrac{d\theta}{dt} = \dfrac{h}{r^2} + (1 - \cos i) \dfrac{d\Omega}{dt}$ (see Art. 29)

$$= \frac{h}{r^2},$$

since $(1 - \cos i)\dfrac{d\Omega}{dt}$, being of the order of the cube of the disturbing force, may be neglected.

Let δr, $\delta\theta$, and δh be corresponding variations in r, θ and h; then

$$\frac{d(\theta + \delta\theta)}{dt} = \frac{h + \delta h}{(r + \delta r)^2},$$

or $\quad \dfrac{d\theta}{dt} + \dfrac{d\delta\theta}{dt} = \dfrac{h}{r^2}\left(1 + \dfrac{\delta h}{h}\right)\left(1 + \dfrac{\delta r}{r}\right)^{-2}$

$$= \frac{h}{r^2} + \frac{\delta h}{r^2} - \frac{2h\delta r}{r^3},$$

neglecting the square of the disturbing force; therefore

$$\frac{d\delta\theta}{dt} = \frac{\delta h}{r^2} - \frac{2h\delta r}{r^3},$$

which gives the variations in longitude. The value of δh may be found from the formula

$$\frac{dh}{dt} = \frac{dR}{d\epsilon} + \frac{dR}{d\varpi}.$$

For the periodic variations in latitude, we refer to Pontécoulant's *Système du Monde*, Tome I. p. 492.

100. As an example of the processes of this Chapter, we will calculate the variations in radius vector and longitude due to the term $m'Me \cos\{(n - 2n')t + \epsilon - 2\epsilon' + \varpi\}$ in R.

Considering this term only, we have

$$R = m'Me \cos\{(n - 2n')t + \epsilon - 2\epsilon' + \varpi\}$$
$$= m'Me \cos\lambda, \text{ suppose.}$$

Hence $\quad \dfrac{dR}{da} = m'\dfrac{dM}{da}e\cos\lambda, \quad \dfrac{dR}{de} = m'M\cos\lambda,$

$$\frac{dR}{d\epsilon} = -m'Me\sin\lambda, \quad \frac{dR}{d\varpi} = -m'Me\sin\lambda,$$

$$\frac{dR}{di} = 0, \quad\quad\quad\quad \frac{dR}{d\Omega} = 0.$$

Substituting these in the formulæ of Art. 59, and neglecting small quantities of orders higher than the first, we have

$$\frac{da}{dt} = -\frac{2na^2}{\mu} m'Me \sin \lambda,$$

$$\frac{de}{dt} = \frac{na}{\mu} m'M \sin \lambda,$$

$$e\frac{d\varpi}{dt} = \frac{na}{\mu} m'M \cos \lambda,$$

$$\frac{d\epsilon}{dt} = -\frac{2na^2}{\mu} m'\frac{dM}{da} e \cos \lambda + \frac{1}{2}\frac{na}{\mu} em'M \cos \lambda,$$

$$\frac{d^2\zeta}{dt^2} = \frac{3n^2a}{\mu} m'Me \sin \lambda.$$

By integration we have

$$\delta a = \frac{2m'M}{\mu}\frac{na^2e}{n-2n'} \cos \lambda,$$

$$\delta \epsilon = -\frac{m'M}{\mu}\frac{na}{n-2n'} \cos \lambda,$$

$$e\delta\varpi = \frac{m'M}{\mu}\frac{na}{n-2n'} \sin \lambda,$$

$$\delta \epsilon = \left(\frac{1}{2}\frac{m'M}{\mu} - \frac{2m'a}{\mu}\frac{dM}{da}\right)\frac{nae}{n-2n'} \sin \lambda,$$

$$\delta\zeta = -\frac{3m'M}{\mu}\frac{n^2ae}{(n-2n')^2} \sin \lambda.$$

Also

$$r = a\left\{1 + \frac{1}{2}e^2 - e \cos(\zeta + \epsilon - \varpi) - \frac{1}{2}e^2 \cos 2(\zeta + \epsilon - \varpi) - \dots\right\};$$

therefore, small quantities of orders higher than the first being. neglected,

$$\frac{dr}{da} = 1 - e \cos (\zeta + \epsilon - \varpi),$$

$$\frac{dr}{de} = a \{e - \cos (\zeta + \epsilon - \varpi) - e \cos 2 (\zeta + \epsilon - \varpi)\},$$

$$\frac{dr}{d\varpi} = - a \{e \sin (\zeta + \epsilon - \varpi) + e^2 \sin 2 (\zeta + \epsilon - \varpi)\},$$

$$\frac{dr}{d\zeta} = ae \sin (\zeta + e - \varpi),$$

$$\frac{dr}{d\epsilon} = ae \sin (\zeta + \epsilon - \varpi).$$

Now $\delta r = \dfrac{dr}{da} \delta a + \dfrac{dr}{de} \delta e + \dfrac{dr}{d\varpi} \delta \varpi + \dfrac{dr}{d\zeta} \delta \zeta + \dfrac{dr}{d\epsilon} \delta \epsilon$

$$= \frac{2m'M}{\mu} \frac{na^2 e}{n - 2n'} \cos \lambda$$

$$- \frac{m'M}{\mu} \frac{na^2}{n - 2n'} \cos \lambda \{e - \cos (\zeta + \epsilon - \varpi) - e \cos 2 (\zeta + \epsilon - \varpi)\}$$

$$- \frac{m'M}{\mu} \frac{na^2}{n - 2n'} \sin \lambda \{\sin (\zeta + \epsilon - \varpi) + \epsilon \sin 2 (\zeta + \epsilon - \varpi)\}.$$

Since we are neglecting the square of the disturbing force, the elements in this equation may be considered as constant, and therefore nt written for ζ: we have then, restoring to λ its value

$$\delta r = \frac{m'M}{\mu} \frac{na^2}{n - 2n'} \cos 2 \{(n - n') t + \epsilon - \epsilon'\}$$

$$+ \frac{m'M}{\mu} \frac{na^2 e}{n - 2n'} \cos \{(n - 2n') t + \epsilon - 2\epsilon' + \varpi\}$$

$$+ \frac{m'M}{\mu} \frac{na^2e}{n-2n'} \cos\{(3n-2n')t + 3\epsilon - 2\epsilon' - \varpi\},$$

which is the variation in radius vector.

101. To calculate the variation in longitude, we shall employ the equation

$$\frac{d\delta\theta}{dt} = \frac{\delta h}{r^2} - \frac{2h\delta r}{r^3}.$$

Now $\dfrac{dh}{dt} = \dfrac{dR}{d\epsilon} + \dfrac{dR}{d\varpi} = -2m'Me\sin\lambda,$

therefore $\delta h = \dfrac{2m'Me}{n-2n'}\cos\lambda;$

therefore $\dfrac{\delta h}{r^2} = \dfrac{2m'Me}{a^2(n-2n')}\cos\lambda$

$$= \frac{2m'M}{\mu}\frac{n^2ae}{n-2n'}\cos\{(n-2n')t + \epsilon - 2\epsilon' + \varpi\}.$$

Also $\dfrac{h\delta r}{r^3} = \dfrac{h\delta r}{a^3}\{1 + 3e\cos(nt + \epsilon - \varpi) + ...\}$

$$= \frac{m'M}{\mu}\frac{n^2a}{n-2n'}\cos 2\{(n-n')t + \epsilon - \epsilon'\}$$

$$+ \frac{5}{2}\frac{m'M}{\mu}\frac{n^2ae}{n-2n'}\cos\{(n-2n')t + \epsilon - 2\epsilon' + \varpi\}$$

$$+ \frac{5}{2}\frac{m'M}{\mu}\frac{n^2ae}{n-2n'}\cos\{(3n-2n')t + 3\epsilon - 2\epsilon' - \varpi\}.$$

Hence by substitution

$$\frac{d\delta\theta}{dt} = -\frac{2m'M}{\mu}\frac{n^2a}{n-2n'}\cos 2\{(n-n')t + \epsilon - \epsilon'\}$$

$$-\frac{3m'M}{\mu}\frac{n^2ae}{n-2n'}\cos\left\{(n-2n')\,t+\epsilon-2\epsilon'+\varpi\right\}$$

$$-\frac{5m'M}{\mu}\frac{n^2ae}{n-2n'}\cos\left\{(3n-2n')\,t+3\epsilon-2\epsilon'-\varpi\right\}.$$

By integration

$$\delta\theta=-\frac{m'M}{\mu}\frac{n^2a}{(n-2n')(n-n')}\sin 2\left\{(n-n')\,t+\epsilon-\epsilon'\right\}$$

$$-\frac{3m'M}{\mu}\frac{n^2ae}{(n-2n')^2}\sin\left\{(n-2n')\,t+\epsilon-2\epsilon'+\varpi\right\}$$

$$-\frac{5m'M}{\mu}\frac{n^2ae}{(n-2n')(3n-2n')}\sin\left\{(3n-2n')t+3\epsilon-2\epsilon'-\varpi\right\},$$

which is the variation in longitude.

In the case of Uranus and Neptune, since $n:n'$ nearly as $2:1$, the term we have been considering is important in the theory of the long inequality.

CHAPTER VII.

DIRECT METHOD OF CALCULATING THE INEQUALITIES IN RADIUS VECTOR, LONGITUDE, AND LATITUDE.

102. In the calculation of the planetary inequalities, we have hitherto employed exclusively the method of the Variation of Elements, but there is another method of solving the problem, which demands our attention. It consists in obtaining equations for calculating the inequalities in radius vector, longitude, and latitude directly from the equations of motion. This method is indeed the simplest to employ in the case of periodic variations of short period, that of the preceding Chapters being the most convenient for the calculation of secular variations and long inequalities. For, since these latter take place with extreme slowness, the elliptic elements, when once corrected for them, continue for a considerable period to represent the actual motion; while, in the former case, the values of the elements change rapidly, and the motion cannot for long be represented by the same ellipse. We proceed, then, to the direct method of calculation*.

103. If r_1, θ_1, and z denote the projected radius vector, longitude, and distance from the plane of reference, of the planet, we have (see Art. 9) the equations of motion

* The two methods are sometimes distinguished as those of Lagrange and Laplace, but in the *Mécanique Céleste* we find both employed.

$$\frac{d^2r_1}{dt^2} - r_1\left(\frac{d\theta_1}{dt}\right)^2 = -\frac{\mu r_1}{r^3} + \frac{dR}{dr_1},$$

$$\frac{d}{dt}\left(r_1^2\frac{d\theta_1}{dt}\right) = \frac{dR}{d\theta_1},$$

$$\frac{d^2z}{dt^2} = -\frac{\mu z}{r^3} + \frac{dR}{dz}.$$

If we take for the fixed plane of reference the position of the plane of the orbit at the epoch from which the time is reckoned, the inclination (as we have remarked in Art. 97) will be the order of the disturbing force, the square of which will be neglected. Now it will be seen on referring to Art. 42, that r_1 and θ_1 differ from r and θ by quantities depending upon the square of the inclination : hence in the above equations, we may replace r_1 and θ_1 by r and θ respectively. Also if λ denote the latitude of the planet, we have

$$z = r \sin \lambda.$$

Hence our equations of motion become

$$\frac{d^2r}{dt^2} - r\left(\frac{d\theta}{dt}\right)^2 = -\frac{\mu}{r^3} + \frac{dR}{dr} \quad\dots\dots\dots\dots\dots(1),$$

$$\frac{d}{dt}\left(r^2\frac{d\theta}{dt}\right) = \frac{dR}{d\theta} \quad\dots\dots\dots\dots\dots\dots(2),$$

$$\frac{d^2 (r \sin \lambda)}{dt^2} = -\frac{\mu}{r^2}\sin\lambda + \frac{dR}{dz} \quad\dots\dots\dots(3).$$

104. As a first approximation, let values of r, θ and λ be obtained from these equations by neglecting the disturbing force, and let $r + \delta r$, $\theta + \delta\theta$, $\lambda + \delta\lambda$ denote the true values of these co-ordinates; then δr, $\delta\theta$ and $\delta\lambda$ will be very small quantities, of the order of the disturbing force : they are termed the *perturbations* in radius vector, longitude, and latitude. We proceed to investigate equations by means of which these quantities may be determined.

105. *To obtain the equation for the perturbation in radius vector.*

From equations (1) and (2) of Art. 103, we obtain

$$\left(\frac{dr}{dt}\right)^2 + r^2\left(\frac{d\theta}{dt}\right)^2 = \frac{2\mu}{r} + 2\int\left(\frac{dR}{dr}\frac{dr}{dt} + \frac{dR}{d\theta}\frac{d\theta}{dt}\right) + C$$

$$= \frac{2\mu}{r} + 2\int\frac{d(R)}{dt}\,dt + C\ldots\ldots\ldots.(4).$$

Multiply (1) by r and add to it (4) : thus

$$r\frac{d^2r}{dt^2} + \left(\frac{dr}{dt}\right)^2 = \frac{\mu}{r} + r\frac{dR}{dr} + 2\int\frac{d(R)}{dt}\,dt + C,$$

i.e. $\dfrac{d^2(r^2)}{dt^2} = \dfrac{2\mu}{r} + 2r\dfrac{dR}{dr} + 4\int\dfrac{d(R)}{dt}\,dt + 2C\ldots\ldots(5).$

If the disturbing force be neglected, this equation becomes

$$\frac{d^2(r^2)}{dt^2} = \frac{2\mu}{r} + 2C\ \ldots\ldots\ldots\ldots\ldots.(6).$$

Let a value of r be obtained from this equation, and let $r + \delta r$ denote the true radius vector : then if we agree to neglect the square of the disturbing force in our next approximation, it will be sufficient to write $r + \delta r$ for r in those terms of (5) which do not involve the disturbing force : also since δr is itself of the order of the disturbing force, its square may be neglected. Hence from (5)

$$\frac{d^2(r+\delta r)^2}{dt^2} = \frac{2\mu}{r+\delta r} + 2r\frac{dR}{dr} + 4\int\frac{d(R)}{dt}\,dt + 2C;$$

therefore $\dfrac{d^2(r^2)}{dt^2} + 2\dfrac{d^2(r\delta r)}{dt^2} = \dfrac{2\mu}{r} - \dfrac{2\mu}{r^3}\delta r + 2r\dfrac{dR}{dr}$

$$+ 4\int\frac{d(R)}{dt}\,dt + 2C;$$

hence by (6),

$$\frac{d^2(r\delta r)}{dt^2} + \frac{\mu}{r^3}(r\delta r) = r\frac{dR}{dr} + 2\int\frac{d(R)}{dt}\,dt,$$

which is the equation for the perturbation in radius vector. We may express the right-hand member in a more convenient form, for since (Art. 42)

$$r = a(1+u),$$

$$\frac{dR}{da} = \frac{dR}{dr}\frac{dr}{da} = (1+u)\frac{dR}{dr};$$

therefore

$$r\frac{dR}{dr} = a\frac{dR}{da};$$

also

$$\frac{d(R)}{dt} = n\frac{dR}{d\epsilon}.$$

Hence our equation becomes

$$\frac{d^2(r\delta r)}{dt^2} + \frac{\mu}{r^3}(r\delta r) = a\frac{dR}{da} + 2n\int\frac{dR}{d\epsilon}\,dt.$$

106. *To obtain the equation for the perturbation in longitude.*

We have from equation (1) of Art. 103,

$$\left(\frac{d\theta}{dt}\right)^2 = \frac{1}{r}\frac{d^2r}{dt^2} + \frac{\mu}{r^3} - \frac{1}{r}\frac{dR}{dr}.$$

As before, let a value of θ be obtained from this equation, the disturbing force being neglected, and let $\theta + \delta\theta$ denote the true value of the longitude: then writing $r + \delta r$ for r and $\theta + \delta\theta$ for θ, and neglecting the square of the disturbing force, we have

$$2\frac{d\theta}{dt}\frac{d\delta\theta}{dt} = \frac{1}{r}\frac{d^2\delta r}{dt^2} - \frac{\delta r}{r^2}\frac{d^2r}{dt^2} - \frac{3\mu}{r^4}\delta r - \frac{1}{r}\frac{dR}{dr}:$$

but

$$r^2\frac{d\theta}{dt} = h, \quad r\frac{dR}{dr} = a\frac{dR}{da};$$

therefore $2h\dfrac{d\delta\theta}{dt} = r\dfrac{d^2\delta r}{dt^2} - \delta r\dfrac{d^2r}{dt^2} - \dfrac{3\mu}{r^3}\delta r - a\dfrac{dR}{da}$

$$= \dfrac{d}{dt}\left(r\dfrac{d\delta r}{dt} - \delta r\dfrac{dr}{dt}\right) - \dfrac{3\mu}{r^3}.r\delta r - a\dfrac{dR}{da}.$$

This equation will become integrable if we eliminate the term $-\dfrac{3\mu}{r^3}.r\delta r$ by means of the equation for the perturbation in radius vector. We have from that equation

$$0 = 3\dfrac{d^2(r\delta r)}{dt^2} + \dfrac{3\mu}{r^3}.r\delta r - 3a\dfrac{dR}{da} - 6n\int\dfrac{dR}{d\epsilon}\,dt\,;$$

therefore by addition,

$$2h\dfrac{d\delta\theta}{dt} = 3\dfrac{d^2(r\delta r)}{dt^2} + \dfrac{d}{dt}\left(r\dfrac{d\delta r}{dt} - \delta r\dfrac{dr}{dt}\right) - 4a\dfrac{dR}{da}$$

$$- 6n\int\dfrac{dR}{d\epsilon}\,dt\,;$$

therefore $2h\delta\theta = 3\dfrac{d(r\delta r)}{dt} + r\dfrac{d\delta r}{dt} - \delta r\dfrac{dr}{dt} - 4a\int\dfrac{dR}{da}\,dt$

$$- 6n\iint\dfrac{dR}{d\epsilon}\,dt^2,$$

the arbitrary constant being considered as included in the sign of integration; therefore

$$2h\delta\theta = 4\dfrac{d(r\delta r)}{dt} - 2\delta r\dfrac{dr}{dt} - 4a\int\dfrac{dR}{da}\,dt - 6n\iint\dfrac{dR}{d\epsilon}\,dt^2,$$

or $\quad h\delta\theta = 2\dfrac{d(r\delta r)}{dt} - \delta r\dfrac{dr}{dt} - 2a\int\dfrac{dR}{da}\,dt - 3n\iint\dfrac{dR}{d\epsilon}\,dt^2,$

which determines the perturbation in longitude, when that in radius vector is known.

C. P. T.

8

107. *To obtain the equation for the perturbation in latitude.*

From equation (3) of Art. 103, we have

$$\frac{d^2(r\sin\lambda)}{dt^2} + \frac{\mu(r\sin\lambda)}{r^3} = \frac{dR}{dz}.$$

Since the plane of reference is supposed to coincide with the position of the plane of the orbit at the epoch from which the time is reckoned, we have at the epoch, $\lambda = 0$: hence, denoting by $\delta\lambda$ the latitude at time t, our equation becomes

$$\frac{d^2(r\delta\lambda)}{dt^2} + \frac{\mu}{r^3}(r\delta\lambda) = \frac{dR}{dz},$$

which is similar in form to the equation for the perturbation in radius vector.

108. *To integrate the equation for the perturbation in radius vector.*

The equation is (Art. 105)

$$\frac{d^2(r\delta r)}{dt^2} + \frac{\mu}{r^3}(r\delta r) = a\frac{dR}{da} + 2n\int\frac{dR}{d\epsilon}\,dt.$$

Let us consider a term in R of the form

$$P\cos\{(pn - qn')t + Q\},$$

where P is a function of the mean distances, excentricities, and inclinations, and Q of the longitudes of the perihelia, nodes, and epochs: then uniting this term with the non-periodic part of R, which we have denoted by $\bar{F}$, we have

$$R = F + P\cos\{(pn - qn')t + Q\};$$

therefore $\quad\dfrac{dR}{da} = \dfrac{dF}{da} + \dfrac{dP}{da}\cos\{(pn - qn')t + Q\},$

$$\frac{dR}{d\epsilon} = -P\frac{dQ}{d\epsilon}\sin\{(pn - qn')t + Q\},$$

since F does not contain ϵ; therefore

$$n \int \frac{dR}{d\epsilon}\, dt = \frac{nP \dfrac{dQ}{d\epsilon}}{pn - qn'} \cos\{(pn - qn')t + Q\} + m'g,$$

where g is an arbitrary constant. It may appear superfluous to introduce this quantity, since an arbitrary constant C has already been added in Art. 105; but its introduction is merely equivalent to a small change in the value of C, and any alteration which does not interfere with the first approximation, obtained by neglecting the disturbing force, is of course permissible. Thus g is a purely arbitrary quantity which might have been omitted, but which we shall find it convenient to retain, leaving its value to be assigned hereafter.

Hence
$$a\frac{dR}{da} + 2n\int \frac{dR}{d\epsilon}\, dt = 2m'g + a\frac{dF}{da}$$

$$+ \left\{ a\frac{dP}{da} + \frac{2nP\dfrac{dQ}{d\epsilon}}{pn - qn'} \right\} \cos\{(pn - qn')t + Q\}$$

$$= 2m'g + a\frac{dF}{da} + P_1 \cos\{(pn - qn')t + Q\},$$

suppose, where
$$P_1 = a\frac{dP}{da} + \frac{2nP\dfrac{dQ}{d\epsilon}}{pn - qn'}.$$

Again (see Art. 13),
$$r = a\left\{ 1 + \frac{1}{2}e^2 - e\cos(nt + \epsilon - \varpi) \right.$$

$$\left. - \frac{1}{2}e^2 \cos 2(nt + \epsilon - \varpi) - \ldots \right\},$$

therefore
$$\frac{\mu}{r^3} = \frac{\mu}{a^3}\left\{ 1 + 3e\cos(nt + \epsilon - \varpi) + \ldots \right\},$$

$$= n^2\left\{ 1 + 3e\cos(nt + \epsilon - \varpi) + \ldots \right\},$$

since
$$n^2 a^3 = \mu.$$

Hence, by substitution, the equation for the perturbation in radius vector becomes

$$\frac{d^2(r\delta r)}{dt^2} + n^2 . r\delta r = 2m'g + a\frac{dF}{da} + P_1\cos\{(pn - qn')t + Q\}$$

$$- n^2 r\delta r\{3e\cos(nt + \epsilon - \varpi) + \ldots\}.$$

109. This equation must be solved by successive approximation, as in the Lunar Theory. By omitting all small quantities, we obtain a first approximation to the value of $r\delta r$; this being substituted in the second member, and small quantities of orders higher than the first neglected, we obtain a second approximation, which will be correct to the first order. In like manner, a third, and higher approximations may be obtained.

On referring to Art. 59, it will be seen that small quantities of the second order being neglected,

$$F = \frac{1}{2}m'C_0;$$

hence, neglecting all small quantities, the equation of the preceding Article becomes

$$\frac{d^2(r\delta r)}{dt^2} + n^2 . r\delta r = 2m'g + \frac{m'}{2}a\frac{dC_0}{da} + P_1\cos\{(pn - qn')t + Q\}.$$

The integral of this equation is

$$r\delta r = \frac{1}{n^2}\left(2m'g + \frac{m'}{2}a\frac{dC_0}{da}\right)$$

$$+ \frac{P_1}{n^2 - (pn - qn')^2}\cos\{(pn - qn')t + Q\}$$

$$+ A\cos(nt - B),$$

where A and B are arbitrary constants. Since, if all small quantities be neglected, $r = a$, we have as a first approximation,

$$\delta r = \frac{1}{n^2 a}\left(2m'g + \frac{m'}{2}\, a\, \frac{dC_0}{da}\right)$$

$$\cdot \quad + \frac{P_1}{a\left\{n^2 - (pn - qn')^2\right\}} \cos\left\{(pn - qn')\,t + Q\right\}$$

$$+ \frac{A}{a}\cos\left(nt - R\right)$$

110. We may, however, omit the last term: for, considering this only, the radius vector of the planet becomes

$$a\left\{1 - e\cos\left(nt + \epsilon - \varpi\right) + \frac{A}{a^2}\cos\left(nt - B\right) + \ldots\right\}$$

$$= a\left[1 - \left\{e\cos\left(\epsilon - \varpi\right) - \frac{A}{a^3}\cos B\right\}\cos nt\right.$$

$$\left. + \left\{e\sin\left(\epsilon - \varpi\right) + \frac{A}{a^2}\sin B\right\}\sin nt + \ldots\right]$$

$$= a\left\{1 - e_1\cos\left(nt + \epsilon - \varpi_1\right) + \ldots\right\},$$

if

$$e_1\cos\left(\epsilon - \varpi_1\right) = e\cos\left(\epsilon - \varpi\right) - \frac{A}{a^3}\cos B,$$

$$e_1\sin\left(\epsilon - \varpi_1\right) = e\sin\left(\epsilon - \varpi\right) + \frac{A}{a^2}\sin B,$$

from which e_1 and ϖ_1 may be determined.

Now since the ellipse upon which our approximations are based, has been obtained by neglecting the disturbing force, we may in the elliptic formulæ replace e and ϖ by e_1 and ϖ_1 respectively, since they differ by quantities of the order of the disturbing force. If this be done, our first approximation becomes

$$\delta r = \frac{1}{n^2 a}\left(2m'g + \frac{m'}{2}\, a\, \frac{dC_0}{da}\right)$$

$$+ \frac{P_1}{a\left\{n^2 - (pn - qn')^2\right\}} \cos\left\{(pn - qn')\,t + Q\right\}.$$

111. In order to obtain a second approximation, this value must be substituted for δr in the right-hand member of the equation of Art. 108. Also since the square of the disturbing force is neglected, we may write e_1 and ϖ_1 for e and ϖ in this equation. We will write for brevity

$$\delta r = L + P_2 \cos \{(pn - qn') t + Q\}.$$

Substituting this in the equation of Art. 108, and omitting those terms which have produced the first approximation*, we have

$$\frac{d^2 . r\delta r}{dt^2} + n^2 . r\delta r$$

$$= - 3n^2 ae_1 \cos (nt + \epsilon - \varpi_1) [L + P_2 \cos \{(pn - qn') t + Q\}]$$

$$= - 3n^2 ae_1 L \cos (nt + \epsilon - \varpi_1)$$

$$- \frac{3}{2} n^2 ae_1 P_2 \cos [\{(p+1) n - qn'\} t + Q + \epsilon - \varpi_1]$$

$$- \frac{3}{2} n^2 ae_1 P_2 \cos [\{(p-1) n - qn'\} t + Q - \epsilon + \varpi_1].$$

112. On the form of this equation, we have an important remark to make. In consequence of the term

$$- 3n^2 ae_1 L \cos (nt + \epsilon - \varpi_1),$$

its integral will contain the term

$$\frac{3}{2} ae_1 nt L \cos (nt + \epsilon - \varpi_1).$$

Thus we are met by a difficulty: our equations have been formed on the hypothesis that the square of δr is small enough to be omitted, whereas here, we have a term capable of indefinite increase. This term, then, if retained, would ultimately vitiate the whole approximation. The difficulty might, as in the Lunar Theory, be obviated by writing

* These terms are omitted for the sake of brevity: in order, therefore, to obtain the complete second approximation, we must add to the integral of the above equation the result of the first approximation.

cn for n in the elliptic formulæ, which amounts to supposing the perihelion to be in motion. Its motion is however better found by the method of the variation of elements. Indeed it may be shewn that such terms as those we are considering lead to the formulæ which have already been obtained, for the secular variation of the elements*. We may accordingly altogether neglect such terms if we suppose the elements of the ellipse on which our approximations are based to have been previously corrected for their secular variations.

113. With this understanding, the complete integral of the equation of Art. 111 will be

$$r\delta r = -\frac{3}{2}\frac{n^2 a e_1 P_2}{n^2 - \{(p+1)n - qn'\}^2}$$
$$\cos[\{(p+1)n - qn'\}t + Q + \epsilon - \varpi_1]$$
$$-\frac{3}{2}\frac{n^2 a e_1 P_2}{n^2 - \{(p-1)n - qn'\}}$$
$$\cos[\{(p-1)n - qn'\}t + Q - \epsilon + \varpi_1]$$
$$+ A\cos(nt - B).$$

If this be added to the result of the first approximation, we obtain for a second approximation

$$r\delta r = \frac{1}{n^2}\left(2m'g + \frac{m'}{2}a\frac{dC_0}{da}\right) + \frac{P_1}{n^2 - (pn - qn')^2}$$
$$\cos\{(pn - qn')t + Q\}$$
$$-\frac{3}{2}\frac{n^2 a e_1 P_2}{n^2 - \{(p+1)n - qn'\}^2}$$
$$\cos[\{(p+1)n - qn'\}t + Q + \epsilon - \varpi_1]$$
$$-\frac{3}{2}\frac{n^2 a e_1 P_2}{n^2 - \{(p-1)n - qn'\}^2}$$
$$\cos[\{(p-1)n - qn'\}t + Q - \epsilon + \varpi_1]$$
$$+ A\cos(nt - B).$$

---

* It is thus that the Secular Variations are first obtained in the *Mécanique Céleste*. See Pontécoulant's *Système du Monde, Supplément au Livre II.*

The arbitrary constants may be disposed of as in the first approximation. In order to obtain a second approximation to the value of δr, it is only necessary to multiply the right-hand member of the above equation by

$$\frac{1}{a}\{1+e_1 \cos (nt + \epsilon - \varpi_1)\},$$

neglecting e_1^2.

114. *To calculate the perturbations in longitude.*

We have (Art. 106)

$$h\delta\theta = 2\frac{d.r\delta r}{dt} - \delta r \frac{dr}{dt} - 2a \int \frac{dR}{da} dt - 3n \iint \frac{dR}{d\epsilon} dt^2.$$

Taking for simplicity, the first approximation to the value of $r\delta r$, which has been obtained by neglecting the first power of the excentricity, we have

$$r\delta r = \frac{1}{n^2}\left(2m'g + \frac{m'}{2} a \frac{dC_0}{da}\right) + \frac{P_1}{n^2 - (pn - qn')^2}$$

$$\cos \{(pn - qn')t + Q\};$$

therefore

$$2\frac{d.r\delta r}{dt} = -\frac{2P_1(pn - qn')}{n^2 - (pn - qn')^2} \sin \{(pn - qn')t + Q\}:$$

also, neglecting the first power of the excentricity,

$$\delta r \frac{dr}{dt} = 0.$$

As in Art. 108, writing $\frac{1}{2} m'C_0$ for F, we have

$$\frac{dR}{da} = \frac{1}{2} m' \frac{dC_0}{da} + \frac{dP}{da} \cos \{(pn - qn')t + Q\},$$

and $\quad n \int \dfrac{dR}{d\epsilon}\, dt = \dfrac{nP\dfrac{dQ}{d\epsilon}}{pn - qn'}\, \cos\{(pn - qn')t + Q\} + m'g;$

therefore $\quad -2a \int \dfrac{dR}{da}\, dt - 3n \iint \dfrac{dR}{d\epsilon}\, dt^2$

$$= f - \left(m'a\, \dfrac{dC_0}{da} + 3m'g\right) t$$

$$- \dfrac{1}{pn - qn'}\left(2a\, \dfrac{dP}{da} + \dfrac{3nP\dfrac{dQ}{d\epsilon}}{pn - qn'}\right) \sin\{(pn - qn')t + Q\},$$

where f is an arbitrary constant.

Hence by substitution, we have

$$h\delta\theta = f - \left(m'a\, \dfrac{dC_0}{da} + 3m'g\right) t$$

$$- \left\{ \dfrac{2a\dfrac{dP}{da}}{pn - qn'} + \dfrac{3nP\dfrac{dQ}{d\epsilon}}{(pn - qn')^2} \right.$$

$$\left. + \dfrac{2P_1(pn - qn')}{n^2 - (pn - qn')^2} \right\} \sin\{(pn - qn')t + Q\}.$$

115. This expression is open to the objection of containing a term proportional to the time, which being capable of indefinite increase, would ultimately vitiate the whole approximation. Here, then, we see the advantage of having a quantity g which may be determined at pleasure: we will so determine it that the objectionable term shall vanish. This condition gives

$$g = -\dfrac{1}{3}\, a\, \dfrac{dC_0}{da}.$$

We may also omit the constant f, and consider it as contained in the epoch. We have then, writing for h its value $na^2 \sqrt{(1 - e^2)}$, and neglecting e^2,

$$\delta\theta = -\frac{1}{na^2}\left\{ \frac{2a\dfrac{dP}{da}}{pn - qn'} + \frac{3nP\dfrac{dQ}{d\epsilon}}{(pn - qn')^2} + \frac{2P_1(pn - qn')}{n^2 - (pn - qn')^2} \right\} \sin\{(pn - qn')t + Q\}.$$

116. Before proceeding to obtain the perturbations in latitude, we will make a few remarks on the forms of the expressions for δr and $\delta\theta$. If we confine ourselves to the results of the first approximation, it will be seen on substituting the value of P_1, that $pn - qn'$ and $n^2 - (pn - qn')^2$ occur as divisors, and that the expression for $\delta\theta$ contains besides, the divisor $(pn - qn')^2$. The second of these may be written

$$\{(1 - p)n + qn'\}\{(1 + p)n - qn'\}.$$

If then either

(i) $pn - qn'$,   (ii) $(1 - p)n + qn'$,   or (iii) $(1 + p)n - qn'$,

be very small, the corresponding terms in δr and $\delta\theta$, though of a high order, may yet be sensible. This is especially the case with the first, since as we have remarked, its square occurs in the expression for $\delta\theta$. These are instances of what in the preceding Chapter have been characterised as *long inequalities*.

The period of the term $P\cos\{(pn - qn')t + Q\}$, which has given rise to these inequalities, is

$$\frac{2\pi}{pn - qn'}:$$

in the case of (i), this is very large, and in that of (ii) or (iii) it is very nearly equal to $\dfrac{2\pi}{n}$, since $pn - qn'$ is nearly equal to $\pm n$. Hence it appears that terms in R whose period is either very large, or nearly equal to that of the planet, may give rise to important inequalities in the radius vector and longitude. Their actual importance will of course depend in part upon the order of the principal part of P with respect to the excentricities and inclinations, i.e. (see Art. 50) upon $p \sim q$.

117. *To integrate the equation for the perturbation in latitude.*

The equation is (Art. 107)

$$\frac{d^2(r\delta\lambda)}{dt^2} + \frac{\mu}{r^3}(r\delta\lambda) = \frac{dR}{dz},$$

the position of the plane of the orbit of the disturbed planet at the epoch being taken for the fixed plane of reference.

Differentiating the expression for R in Art. 44, with respect to z, we obtain

$$- m'(z - z')\left(\tfrac{1}{2}D_0 + D_1\cos\phi + \ldots + D_k\cos k\phi + \ldots\right) - \frac{m'z'}{a'^3},$$

or, putting z equal to 0, and substituting

$$a'\tan i''\sin(n't + \epsilon' - \Omega')$$

for z' (see Art. 42),

$$m'a'\tan i''\sin(n't + \epsilon' - \Omega')\left\{\tfrac{1}{2}D_0 - \frac{1}{a'^3} + D_1\cos\phi + \ldots\right\}.$$

This expression, after reduction, consists of terms of the form

$$P\sin\{(pn - qn')t + Q\},$$

where p and q are positive integers, and either may be zero. Considering one such term, our equation becomes

$$\frac{d^2(r\delta\lambda)}{dt^2} + \frac{\mu}{r^3}(r\delta\lambda) = P\sin\{(pn - qn')t + Q\}.$$

Now as in Art. 108,

$$\frac{\mu}{r^3} = n^2 \{1 + 3e \cos (nt + \epsilon - \varpi) + \ldots\};$$

hence, neglecting the product $e\delta\lambda$, we have for a first approximation

$$\frac{d^2 (r\delta\lambda)}{dt^2} + n^2 . r\delta\lambda = P \sin \{(pn - qn') t + Q\}.$$

The integral of this equation is

$$r\delta\lambda = \frac{P}{n^2 - (pn - qn')^2} \sin \{(pn - qn') t + Q\} + A \cos (nt - B).$$

If instead of taking for the fixed plane of reference, the plane of the orbit of the disturbed planet at epoch, we take a plane slightly inclined to this, we may omit the arbitrary term. For, denoting the planet's latitude with respect to this plane by λ, we have approximately

$$\lambda = \tan i \sin (nt + \epsilon - \Omega),$$

and it may be shewn as in Art. 110, that omitting the term in question is only equivalent to changing slightly the values of i and Ω.

CHAPTER VIII.

118. In the preceding Chapters, we have supposed the planetary motions to take place in free space, and the results of calculations based upon this hypothesis manifest a very close agreement with observation. There is, however, a remarkable circumstance connected with Encke's comet which seems to indicate the possibility of the existence of a very rare medium, too rare indeed to cause any sensible resistance to the motions of the planets, but which, as we shall presently see, may yet influence the motions of comets, in consequence of the extreme smallness of the masses of these bodies. It has been observed that the comet above referred to (which describes an elliptic orbit in a period of about $3\frac{1}{4}$ years,) has since its appearance in 1786, been moving round the Sun with an increasing mean motion. Encke attributes this to the resistance of a medium pervading space. We shall therefore proceed to examine the effects which such a medium would produce upon the elements of a planet's orbit, assuming, in accordance with the usual theory, that the resistance varies as the product of the density of the medium and the square of the velocity of the planet. We shall neglect, in the present investigation, all forces except the Sun's attraction and the resistance of the medium; consequently the planet may be supposed to move wholly in one plane.

119.   Let r, θ be the radius vector and longitude of the planet, s the length of an arc of its actual orbit measured from some fixed point to its position at time t, and ρ the density of the medium.   Then if k be a constant, we may represent the resistance on the planet by $k\rho \left(\dfrac{ds}{dt}\right)^2$, and the equations of motion will be

$$\frac{d^2r}{dt^2} - r\left(\frac{d\theta}{dt}\right)^2 = -\frac{\mu}{r^2} - k\rho \left(\frac{ds}{dt}\right)^2 \frac{dr}{ds},$$

$$\frac{1}{r}\frac{d}{dt}\left(r^2\frac{d\theta}{dt}\right) = -k\rho\left(\frac{ds}{dt}\right)^2 r\frac{d\theta}{ds}.$$

If $r^2\dfrac{d\theta}{dt} = h$, these may be written

$$\frac{d^2r}{dt^2} - r\left(\frac{d\theta}{dt}\right)^2 = -\frac{\mu}{r^2} - k\rho \frac{ds}{dt}\frac{dr}{dt}\dots\dots\dots\dots(1),$$

$$\frac{d}{dt}\left(r^2\frac{d\theta}{dt}\right) = -k\rho h \frac{ds}{dt}\dots\dots\dots\dots\dots(2).$$

These equations are the same in form with those of Art. 20, and may be treated in a similar manner, $-k\rho \dfrac{ds}{dt}\dfrac{dr}{dt}$ taking the place of $\dfrac{dR}{dr}$, and $-k\rho h\dfrac{ds}{dt}$ that of $\dfrac{dR}{d\theta}$.   (See Art. 24.)   We have from equation (2)

$$\frac{dh}{dt} = -k\rho h \frac{ds}{dt}.$$

120.   *To obtain a formula for calculating the mean distance.*

We might proceed as in Art. 25, but we shall here employ the method of Art. 26.   We have

$$\frac{d^2s}{dt^2} = -\frac{\mu}{r^2}\frac{dr}{ds} - k\rho\left(\frac{ds}{dt}\right)^2,$$

and by a known formula of elliptic motion

$$\left(\frac{ds}{dt}\right)^2 = \frac{2\mu}{r} - \frac{\mu}{a}.$$

Differentiating the latter, we obtain

$$2\frac{ds}{dt}\frac{d^2s}{dt^2} = -\frac{2\mu}{r^2}\frac{dr}{dt} + \frac{\mu}{a^2}\frac{da}{dt},$$

and multiplying the former by $2\dfrac{ds}{dt}$,

$$2\frac{ds}{dt}\frac{d^2s}{dt^2} = -\frac{2\mu}{r^2}\frac{dr}{dt} - 2k\rho\left(\frac{ds}{dt}\right)^3;$$

therefore

$$\frac{\mu}{a^2}\frac{da}{dt} = -2k\rho\left(\frac{ds}{dt}\right)^3,$$

or

$$\frac{da}{dt} = -\frac{2k\rho a^2}{\mu}\left(\frac{ds}{dt}\right)^3.$$

121.   *To obtain a formula for calculating the excentricity.*

We have, as in Art. 27,

$$\frac{\mu^2 e^2}{h^2} = \left(\frac{dr}{dt}\right)^2 + \left(\frac{h}{r} - \frac{\mu}{h}\right)^2 \ldots\ldots\ldots\ldots\ldots(3).$$

Differentiating as if r were constant, and writing $-k\dfrac{dr}{dt}\dfrac{ds}{dt}$ for $\dfrac{d^2r}{dt^2}$,

$$\frac{\mu^2 e}{h^2}\frac{de}{dt} = -k\rho\left(\frac{dr}{dt}\right)^2\frac{ds}{dt} + \left\{\left(\frac{h}{r} - \frac{\mu}{h}\right)\left(\frac{1}{r} + \frac{\mu}{h^2}\right) + \frac{\mu^2 e^2}{h^3}\right\}\frac{dh}{dt}$$

$$= -k\rho\left(\frac{dr}{dt}\right)^2\frac{ds}{dt} - \left\{\frac{h}{r^2} - \frac{\mu^2(1-e^2)}{h^3}\right\}k\rho h\frac{ds}{dt}$$

$$= -k\rho\frac{ds}{dt}\left\{\left(\frac{dr}{dt}\right)^2 + \frac{h^2}{r^2} - \frac{\mu^2(1-e^2)}{h^2}\right\}.$$

Now from equation (3)

$$\left(\frac{dr}{dt}\right)^2 + \frac{h^2}{r^2} = \frac{\mu^2 e^2}{h^2} + \frac{2\mu}{r} - \frac{\mu^2}{h^2}$$

$$= \frac{2\mu}{r} - \frac{\mu^2(1-e^2)}{h^2};$$

therefore

$$\frac{\mu^2 e}{h^2}\frac{de}{dt} = -2k\rho\frac{ds}{dt}\left\{\frac{\mu}{r} - \frac{\mu^2(1-e^2)}{h^2}\right\};$$

or

$$\frac{de}{dt} = -\frac{2k\rho}{e}\frac{ds}{dt}\left\{\frac{h^2}{\mu r} - (1-e^2)\right\}$$

$$= -\frac{2k\rho(1-e^2)}{e}\left(\frac{a}{r} - 1\right)\frac{ds}{dt}.$$

This result may also be obtained by differentiating the formula $h^2 = \mu a(1-e^2)$, and substituting the expressions for $\frac{dh}{dt}$ and $\frac{da}{dt}$, as in Art. 28.

122. *To obtain a formula for calculating the longitude of perihelion.*

We have, as in Art. 29,

$$\frac{dr}{dt}\cot(\theta - \varpi) = \frac{h}{r} - \frac{\mu}{h}\dots\dots\dots\dots\dots(4).$$

Differentiating as if r and θ were constant, and writing $-k\rho\frac{dr}{dt}\frac{ds}{dt}$ for $\frac{d^2r}{dt^2}$,

$$\frac{dr}{dt}\operatorname{cosec}^2(\theta - \varpi)\frac{d\varpi}{dt} - k\rho\frac{dr}{dt}\frac{ds}{dt}\cot(\theta - \varpi) = \left(\frac{1}{r} + \frac{\mu}{h^2}\right)\frac{dh}{dt},$$

or since

$$\frac{dh}{dt} = -k\rho h\frac{ds}{dt},$$

$$\frac{dr}{dt} \operatorname{cosec}^2 (\theta - \varpi) \frac{d\varpi}{dt} = k\rho \frac{ds}{dt} \left\{ \frac{dr}{dt} \cot (\theta - \varpi) - \left(\frac{h}{r} + \frac{\mu}{h} \right) \right\}$$

$$= -2 \frac{k\rho\mu}{h} \frac{ds}{dt}, \text{ by (4)};$$

but from equation (4) of Art. 22,

$$\frac{dr}{dt} \operatorname{cosec} (\theta - \varpi) = \frac{\mu e}{h};$$

therefore $\qquad \dfrac{d\varpi}{dt} = -\dfrac{2k\rho}{e} \sin (\theta - \varpi) \dfrac{ds}{dt}.$

123. *To obtain a formula for calculating the longitude of the epoch.*

Wo have (see Art. 13),

$$\theta = \varpi + f (nt + \epsilon - \varpi, e).$$

Differentiating, the elements being considered variable,

$$\frac{d\theta}{dt} = \frac{d\varpi}{dt} + \frac{1}{n} \left(\frac{df}{dt} \right) \frac{d (nt + \epsilon - \varpi)}{dt} + \frac{df}{de} \frac{de}{dt};$$

and differentiating as if the elements were invariable,

$$\frac{d\theta}{dt} = \left(\frac{df}{dt} \right).$$

Equating the two values of $\dfrac{d\theta}{dt}$, we have

$$\frac{d\varpi}{dt} + \frac{1}{n} \frac{d\theta}{dt} \left(t \frac{dn}{dt} + \frac{de}{dt} - \frac{d\varpi}{dt} \right) + \frac{df}{de} \frac{de}{dt} = 0.$$

But $\dfrac{d\theta}{dt} = \dfrac{h}{r^2}$, $\dfrac{df}{de} = \dfrac{d\theta}{de} = a \left(\dfrac{\mu}{h^2} + \dfrac{1}{r} \right) \sin (\theta - \varpi)$ (Art. 17);

therefore $\qquad \dfrac{de}{dt} = -t \dfrac{dn}{dt} + \left(1 - \dfrac{nr^2}{h} \right) \dfrac{d\varpi}{dt}$

$$- \frac{nar^2}{h} \left(\frac{\mu}{h^2} + \frac{1}{r} \right) \sin (\theta - \varpi) \frac{de}{dt}.$$

As in Art. 37, we may omit the term $t \dfrac{dn}{dt}$, if we bear in mind that the mean longitude will then be denoted by $\int n dt + \epsilon$. Thus, on substituting the values of $\dfrac{d\varpi}{dt}$ and $\dfrac{de}{dt}$, we obtain

$$\frac{d\epsilon}{dt} = -\frac{2k\rho}{e} \sin(\theta - \varpi) \left\{ 1 - \frac{nr^2}{h} \right.$$

$$\left. -\frac{nar^2}{h} \left(\frac{\mu}{h^2} + \frac{1}{r} \right) \left(\frac{a}{r} - 1 \right) (1 - e^2) \right\} \frac{ds}{dt} \; .$$

124. We shall now express the results of the preceding Articles in a form convenient for application; and, for simplicity, terms of the order of the square of the disturbing force will be neglected.

If u denote the excentric anomaly, we have

$$r = a(1 - e \cos u) \dots\dots\dots\dots\dots(1),$$

$$nt + \epsilon - \varpi = u - e \sin u \dots\dots\dots(2).$$

Hence
$$\left(\frac{ds}{dt} \right)^2 = \frac{2\mu}{r} - \frac{\mu}{a}$$

$$= \frac{\mu}{a} \left(\frac{2}{1 - e \cos u} - 1 \right)$$

$$= \frac{\mu}{a} \frac{1 + e \cos u}{1 - e \cos u} \; ;$$

therefore $\dfrac{ds}{dt} = na \sqrt{\left(\dfrac{1 + e \cos u}{1 - e \cos u} \right)}$, since $na^{\frac{3}{2}} = \sqrt{\mu}$.

And to transform the independent variable from t to u, we have from equation (2)

$$\frac{dt}{du} = \frac{1}{n}(1 - e \cos u) - \frac{1}{n} \frac{d\epsilon}{du} + \frac{1}{n} \frac{d\varpi}{du} \; ,$$

of which the two last terms, being of the order of the disturbing force, may be omitted, since the terms in which the substitution is to be made are themselves of the first order.

To obtain an expression for $\sin(\theta - \varpi)$, which occurs in the formulæ for the longitudes of perihelion and of the epoch, we have from equation (1), omitting terms of the order of the disturbing force,

$$\frac{dr}{dt} = ae \sin u \, \frac{du}{dt};$$

and from the equation

$$\frac{1}{r} = \frac{\mu}{h^2}\{1 + e\cos(\theta - \varpi)\} \quad \text{(Art. 22),}$$

$$\frac{dr}{dt} = \frac{\mu e}{h}\sin(\theta - \varpi).$$

Equating the two values of $\dfrac{dr}{dt}$,

$$\sin(\theta - \varpi) = \frac{ha}{\mu}\sin u \, \frac{du}{dt}.$$

Hence, by substitution, the formulæ of the preceding Articles become

$$\frac{da}{du} = -2k\rho a^2 (1 + e\cos u)\sqrt{\left(\frac{1 + e\cos u}{1 - e\cos u}\right)},$$

$$\frac{de}{du} = -2k\rho a (1 - e^2)\cos u \sqrt{\left(\frac{1 + e\cos u}{1 - e\cos u}\right)},$$

$$\frac{d\varpi}{du} = -\frac{2k\rho a \sqrt{(1 - e^2)}}{e}\sin u \sqrt{\left(\frac{1 + e\cos u}{1 - e\cos u}\right)},$$

$$\frac{d\epsilon}{du} = \frac{2k\rho a}{e}\{1 - \sqrt{(1 - e^2)} - e^2\cos u\}\sin u \sqrt{\left(\frac{1 + e\cos u}{1 - e\cos u}\right)}.$$

9—2

125. These formulæ are sufficient to determine the elements of the orbit at any time, and being perfectly general, are applicable as well to the motion of comets, as to that of planets, but before we can integrate them, we shall require a knowledge of the form of ρ. Now the analogy of the terrestrial atmosphere would lead us to suppose that if the Sun be surrounded by an ethereal medium, its density decreases as the distance from the Sun increases. Moreover, the researches of Professor Encke on the comet which bears his name, seem to indicate the law of the inverse square. We will, however, merely assume ρ to be such a function of r, that when multiplied by $\sqrt{\left(\dfrac{1 + e \cos u}{1 - e \cos u}\right)}$, and developed in a series of cosines of u and its multiples, it takes the form

$$A + Be \cos u + Ce^2 \cos 2u + \ldots$$

Thus our formulæ become

$$\frac{da}{du} = -2ka^2 \{A + (A + B) e \cos u + \ldots\},$$

$$\frac{de}{du} = -2ka \left\{A \cos u + \frac{Be}{2} (1 + \cos 2u) + \ldots\right\},$$

$$e \frac{d\varpi}{du} = -2ka \left\{A \sin u + \frac{Be}{2} \sin 2u + \ldots\right\},$$

$$\frac{d\epsilon}{du} = ka (Ae \sin u + \ldots).$$

126. *Supposing the orbit nearly circular, to examine the effects of the medium upon the elements of the orbit.*

Since the orbit is nearly circular, we shall neglect squares and higher powers of e; thus the preceding formulæ give on integration

$$a = \text{const.} - 2ka^2 \{Au + (A + B) \, e \sin u\},$$

$$e = \text{const.} - 2ka \left\{ A \sin u + \frac{Be}{2} \left(u + \frac{\sin 2u}{2} \right) \right\},$$

$$e\varpi = \text{const.} + 2ka \left\{ A \cos u + \frac{Be}{4} \cos 2u \right\},$$

$$= \text{const.} - kaAe \cos u.$$

Hence in an entire revolution of the planet, the mean distance is diminished by $4\pi ka^2 A$, and the excentricity by $2\pi kaBe$, while the longitudes of perihelion, and of the epoch, remain unchanged. Also from the formula $n = \dfrac{\sqrt{\mu}}{a^{\frac{3}{2}}}$, it appears that the mean motion is, in an entire revolution, increased by $6\pi knaA$.

127. We have already remarked that no traces of a resisting medium have yet been discovered in the motion of the planets: but, since k varies inversely as the mass of the body acted upon, the formulæ of Art. 124 shew that such a medium, though too rare to influence the planets, might yet sensibly affect the motions of comets, in consequence of the extreme smallness of their masses.

PROBLEMS.

1. SUPPOSING in the Problem of the Three Bodies the relative orbit of two of the bodies to be a circle described uniformly, obtain equations for determining the motion of the third body; and transform the system of co-ordinates, so that the plane of the circular orbit being that of xy, the axis of x shall always pass through the two bodies in that plane.

2. Shew that the plane of the orbit of a planet revolves about the planet's radius vector as an instantaneous axis*.

3. A particle is describing an orbit round a centre of force which is any function of the distance, and is acted upon by a disturbing force which is always perpendicular to the plane of the instantaneous orbit, and inversely proportional to the distance of the body from the centre of the principal force. Prove that the plane of the instantaneous orbit revolves uniformly round its instantaneous axis.

4. A particle, acted on by a force varying as the inverse square of the distance and always tending to a fixed point, suffers slight disturbances: prove that generally there is a conic with the fixed point for focus, with which the body's actual path has a closer contact than with the instantaneous conic.

* In this and the following problem, the plane of the orbit must be supposed to have no angular velocity about a normal to itself. See note to Art. 19.

5. Find when the curvature of the instantaneous orbit of a body, acted on by disturbing forces, is the same as that of the actual orbit; and shew that this is always the case when the only disturbing force arises from the action of a resisting medium.

6. If R be expressed on the one hand as a function of r_1, θ_1, and z (Art. 11), and on the other as a function of r, θ, i, and Ω, θ being measured on the plane of the orbit *from the node*, prove that

$$\frac{dR}{d\theta_1} = \frac{dR}{d\Omega},$$

and obtain a formula for calculating the inclination.

7. If R be expressed as a function of t and the usual elements, obtain the formulæ

$$\frac{d\Omega}{dt} = \frac{1}{h \sin i}\frac{dR}{di}, \quad \frac{di}{dt} = \frac{\cot(\theta-\Omega)}{h}\frac{dR}{di};$$

where θ is measured on the plane of reference as far as the node, and thence on that of the orbit, and

$$h = r^2\left(\frac{d\theta}{dt} - 2\sin^2\frac{i}{2}\frac{d\Omega}{dt}\right).$$

8. The central force being $\frac{\mu}{r^2} + \frac{\mu'}{r^4}$, obtain the following equation for the apsidal motion,

$$\frac{d\varpi}{dt} = \frac{\sqrt{a(1-e^2)}}{e\sqrt{\mu}}\frac{\mu'\cos(\theta-\varpi)}{r^4},$$

a, e and ϖ being elements of the instantaneous ellipse.

9. If the central force on a planet be $\frac{\mu}{r^2} + \mu'$, shew that the planet may be supposed at any instant to be moving in an

ellipse, of which the mean distance is inversely proportional to the planet's distance from the Sun, provided $\frac{\mu}{\mu'}=2\rho\alpha$, where ρ and α are two simultaneous values of r and a. Shew also that the *latus rectum* of this ellipse is constant.

10. When the disturbing function R is independent of θ, find expressions for $\frac{de}{dt}$ and $\frac{d\varpi}{dt}$.

If $R=\frac{m'}{r}$, these expressions give variable values for e and ϖ, whereas the motion of the body actually takes place in a fixed ellipse: shew. this, and explain the apparent paradox.

11. A planet describes an orbit under the action of a force $\frac{\mu}{r^2}$ tending to the Sun, μ not being quite constant: obtain the following equations for the variations of the excentricity and longitude of perihelion:

$$\frac{d(\mu e)}{d\mu}=-\cos(\theta-\varpi),$$

$$\mu e\frac{d\varpi}{d\mu}=-\sin(\theta-\varpi).$$

If $d\mu$ be always positive, what in a whole revolution is the nature of its effects upon the excentricity and position of the major axis?

12.   Prove that for any closed central orbit, the element introduced in the equation of *vis viva* is subject to periodic variations only, when the disturbing force is due to forces tending to centres and functions of the distances from those centres.

13. If the equation of the Moon's orbit be reduced to the form

$$\frac{d^2u}{d\theta^2} + u - a = af,$$

shew that the excentricity and longitude of perihelion may be found from the equations

$$\frac{de}{d\theta} = -f \sin (\theta - \varpi), \quad e\frac{d\varpi}{d\theta} = f\cos (\theta - \varpi).$$

Apply these equations to find e and ϖ, when f is a small disturbing force, depending only upon the Moon's distance from the Earth.

14. Assuming the differential equation for s in the Lunar Theory to be

$$\frac{d^2s}{d\theta^2} + s = -m^2s\left\{\frac{3}{2} + \frac{3}{2}\cos 2(\theta - m\theta)\right\}$$

$$+ m^2\frac{ds}{d\theta}\left\{\frac{3}{2}\sin 2(\theta - m\theta)\right\},$$

shew that if γ be the longitude of the Moon's node,

$$\frac{d\gamma}{d\theta} = -\frac{3}{4}m^2\left\{1 - \cos 2(m\theta - \gamma) - \cos 2(\theta - \gamma)\right.$$

$$\left. + \cos 2(\theta - m\theta)\right\}.$$

From the above expression for $\frac{d\gamma}{d\theta}$, find the ratio of the mean motion of the node to that of the Moon, taking into account terms of the order m^4.

15. If two planets disturbing one another were revolving in periods of 350 and 201 days, what form of terms in the disturbing function would demand examination?

16. The periods of Venus and the Earth are 224·7 and 365·256 days respectively; find approximately the period of

the long inequality arising from their mutual perturbations, the important term in the disturbing function R being of the form

$$Pe^{3}e'^{2} \cos \{13 \, (nt + \epsilon) - 8 \, (n't + \epsilon') - 3\varpi - 2\varpi'\}.$$

17. The radius vector of a planet is affected with a small periodical inequality; shew that its effect may be represented by continued and periodical alterations of the excentricity and longitude of perihelion, the period of either being $\dfrac{PT}{P \sim T}$, where P is the period of the planet and T that of the inequality.

18. If in addition to the force of the Sun on a planet there be a small force tending towards the Sun, and varying inversely as the mth power of the distance of the planet from the Sun, prove that the perihelion of the orbit will have a progressive or regressive motion according as m is greater or less than 2.

Can you explain this result by reasoning similar to that used in Airy's *Gravitation?*

19. It has been found by comparing theory with observation that the perihelion of Mercury progresses at a rate greater by α than that due to the attraction of known bodies: shew that this increment would be accounted for if the law of force tending to the Sun were $\dfrac{\mu}{r^{2}} + \dfrac{\mu'}{r^{4}}$, and if $\mu' = ac^{4}\sqrt{\dfrac{\mu}{c}}$, the orbit being supposed to be nearly a circle, and the mean distance to be c.

20. The central force acting on a body being

$$\frac{\mu}{r^{2}} + \mu'\phi \, (r),$$

shew to terms inclusive of μ' and the square of the excentricity, that the motion is in an ellipse revolving uniformly about the focus.

21. Shew by means of the formula

$$\frac{da}{dt} = \frac{2na^2}{\mu}\frac{dR}{d\epsilon},$$

that the chief perturbation of the axis major of the Moon's orbit may be expressed by the equation

$$a_1 = a\left\{1 + \frac{3n'^2}{2n(n-n')}\cos 2(nt+\epsilon - n't - \epsilon')\right\},$$

where n and n' are the mean motions of the Moon and Sun respectively.

22. A satellite revolving in an ellipse of small excentricity is disturbed by another satellite revolving about the same primary; find approximately the variation of the mean distance and the motion of the apse, corresponding to the terms

$$\frac{n'^2}{4}r^2[1 + 3\cos\{2(n-n')t+\epsilon - \epsilon'\}]$$

in the function R, having given

$$\frac{da}{dt} = \frac{2na^2}{\mu}\frac{dR}{d\epsilon}, \quad \frac{d\varpi}{dt} = \frac{na\sqrt{(1-e^2)}}{\mu e}\frac{dR}{de}.$$

23. Prove that, neglecting periodical variations, the excentricity of any orbit can always be represented by the diagonal of a parallelogram, whose sides are constant, and angle varies uniformly.

24. Given the equations

$$\tan^2 i = N_1^2 + N_2^2 + 2N_1N_2\cos(h_1 t + \delta_1 - \delta_2),$$

$$\tan\Omega = \frac{N_1\sin(h_1 t + \delta_1) + N_2\sin\delta_2}{N_1\cos(h_1 t + \delta_1) + N_2\cos\delta_2};$$

explain the nature of the motion of the node, when the minimum inclination is zero.

25.   Prove that as far as secular variations only are concerned the function F is constant.

26.   Considering only secular variations, obtain the following equations:

$$\Sigma\left(m\sqrt{a}\,e^2\frac{d\varpi}{dt}\right)=C,\quad \Sigma\left(m\sqrt{a}\tan^2 i\frac{d\Omega}{dt}\right)=C.$$

27.   If the squares of the masses of two mutually disturbing planets were to each other inversely as their mean
distances, shew that the nodes would oscillate through equal
angles.

28.   If M, m, m' be the masses of three bodies mutually
attracting according to the law of gravity, M being much
larger than m or m', and if v, v' be the velocities of m, m' at
distances r, r' from the centre of M, supposed fixed, shew that
the equation of *vis viva* for this case may be assumed to be

$$mv^2+m'v'^2+2M\left(\frac{m}{2a}-\frac{m}{r}+\frac{m'}{2a'}-\frac{m'}{r'}\right)=0,$$

$2a$ and $2a'$ being the major axes of the instantaneous ellipses
of m and m'.

29.   Infer from the foregoing equation by the method of
the variation of parameters the ratio of simultaneous changes
in the mean distances and mean motions of two planets
mutually disturbing.

30.   If r be the true radius vector, θ, the projected longitude, and λ the latitude of a planet, obtain the following
equation of motion:

$$\frac{d^2r}{dt^2}-r\cos^2\lambda\left(\frac{d\theta_1}{dt}\right)^2-r\left(\frac{d\lambda}{dt}\right)^2+\frac{\mu}{r^2}=\frac{dR}{dr}.$$

31. Obtain the following equation between the perturbations of a planet in longitude and radius vector, whatever be the law of force, provided it be central and a function of the distance only, and provided such a function as R can be found:

$$h . \delta\theta = \frac{d}{dt} (2r \, \delta r) - \frac{1}{r} \frac{dr}{dt} r \delta r + 3 \int \frac{d(R)}{dt} dt + 2r \frac{dR}{dr}$$

$$- 4F \delta r - 2r \frac{dF}{dr} \delta r,$$

where F denotes the central force, and h twice the sectorial area described by the undisturbed planet round the Sun.

32. If the orbits of two planets which disturb each other be very nearly circular, shew that the inequalities of the radius vector may be immediately deduced from those of the longitude by means of the equation

$$\frac{\delta r}{r} + \frac{1}{2} \frac{d . \delta\theta}{\frac{dt}{dt}} - \frac{1}{2} \frac{a}{\mu} \frac{n}{n - n'} R = 0.$$

33. Integrate the equation

$$\frac{d^2 (r \delta r)}{dt^2} + n^2 . r \delta r = \Sigma \{ P \cos (pnt + Q) \},$$

determining the arbitrary constants so that $\delta r = 0$, and $\frac{d . \delta r}{dt} = 0$, when $t = 0$: and shew that for small values of t,

$$r \delta r = \Sigma \left(\frac{Pt^2}{2} \cos Q \right),$$

the case of $p = 1$ being included.

34. A planet moves in a resisting medium of which the resistance

$$= \frac{f}{r^2} \left(\frac{ds}{dt} \right)^2 ;$$

apply the equation

$$\frac{d^2 (r\delta r)}{dt^2} + \frac{\mu}{r^3} \cdot r\delta r - 2 \int \frac{d(R)}{dt} dt - r \frac{dR}{dr} = 0$$

to obtain the following, in which e^2 is neglected :

$$\frac{d^2 (r\delta r)}{dt^2} + n^2 \cdot r\delta r + n^2 \cdot r\delta r \cdot 3e \cos (nt + \epsilon - \varpi)$$

$$+ 2fn^2 a \left\{ nt + \frac{11}{2} e \sin (nt + \epsilon - \varpi) \right\} = 0.$$

35. The co-ordinates of the position at any time t of a disturbed planet being $x + \delta x$, $y + \delta y$, $z + \delta z$, reckoned from the Sun's centre as a fixed origin, and referred to the plane of motion at a given epoch; and r being the heliocentric distance, x, y the co-ordinates of the position which the planet would have had at the time t, if the disturbance had ceased at the given epoch; obtain the following equations for determining δx, δy, δz to the first order of the disturbing force :

$$\frac{d^2 (\delta x)}{dt^3} + \frac{\mu}{r^3} \left\{ \left(1 - \frac{3x^2}{r^2} \right) \delta x - \frac{3xy}{r^2} \delta y \right\} + \frac{dR'}{dx'} = 0,$$

$$\frac{d^2 (\delta y)}{dt^2} + \frac{\mu}{r^3} \left\{ \left(1 - \frac{3y^2}{r^2} \right) \delta y - \frac{3xy}{r^2} \delta x \right\} + \frac{dR'}{dy'} = 0,$$

$$\frac{d^2 (\delta z)}{dt^2} + \frac{\mu \delta z}{r^3} + \frac{dR'}{dz'} = 0,$$

in which μ is the sum of the masses of the Sun and planet, and R' is put for

$$- m' (x'^2 + y'^2 + z'^2)^{-\frac{1}{2}} + m' \left\{ (x - x')^2 + (y - y')^2 + (z - z')^2 \right\}^{-\frac{1}{2}},$$

m' being the mass, and x', y', z' the heliocentric co-ordinates of the disturbing planet.

36. Shew that the effect of a resisting medium on the instantaneous orbit of a planet, would be to make the apsidal line regrede or progrede, according as the planet moved from perihelion to aphelion, or from aphelion to perihelion.

37. Two small planets P, Q, very near each other, revolve about the Sun in orbits very nearly circular, and make two revolutions about each other while they make one revolution about the Sun. Compare the sum of their masses with the mass of the Sun.

If the line PQ move parallel to itself, what inference do you draw?

38. If the motion of a planet round the Sun be disturbed by the action of another planet, the latter being supposed to describe a circular orbit of radius a' with uniform velocity n, obtain the following exact equation:

$$\left(\frac{dr_1}{dt}\right)^2 + r_1^2\left(\frac{d\theta_1}{dt}\right)^2 + \left(\frac{dz}{dt}\right)^2 - 2nr_1^2\frac{d\theta_1}{dt} + C$$

$$= \frac{2\mu}{r} - \frac{2m'r}{a'^2}\cos\omega + \frac{2m'}{(a'^2 - 2a'r\cos\omega + r^2)^{\frac{1}{2}}},$$

where r is the radius vector of the disturbed planet, r_1, θ_1, z its co-ordinates referred to a fixed plane, and ω the inclination of the radii vectores of the disturbed and disturbing planets to each other.

39. Prove that, if the periodic times of a disturbed and disturbing planet are not commensurable, the secular changes of the orbit of the disturbed planet are the same as they would be if the mass of the disturbing planet were distributed over its orbit, in such a manner, that the part of the mass distributed over each portion of the orbit should be proportional to the time which the planet actually takes to describe that portion.

Additional Examples.

40. If the disturbing force be radial and equal to $\mu'r$, prove that the mean distance at any time is

$$a - 2a\, \frac{a^3\mu'}{\mu} \left\{ e \cos\left(nt + \epsilon - \varpi\right) + \frac{e^2}{4} \cos 2\left(nt + \epsilon - \varpi\right) \right\}.$$

(*Math. Tripos*, 1872.)

41. A particle moves in a plane subject to a force directed to a fixed point in the plane, and depending on the distance from the point: prove these equations:

$$\frac{h^2}{a^3}\frac{de}{dt} + \frac{h}{b^2}\frac{db}{dt} = -\frac{dR}{da},$$

$$\frac{h^2}{b^3}\frac{de}{dt} - \frac{h}{b^2}\frac{da}{dt} = -\frac{dR}{db},$$

$$a\frac{de}{dt} - (t + \epsilon)\frac{da}{dt} = 0;$$

where a and b are the intercepts of the direction of motion on fixed axes through the fixed point, h is an absolute constant, R is the potential of the force and ϵ a quantity such that, v being the velocity of the particle at time t, $v(t + \epsilon)$ is its distance measured along the direction of motion from the axis of x.   (*M. T.* 1873.)

42. A large number of meteors are distributed uniformly in the circumference of a circle of radius b concentric with the sun and in the plane of the planet's orbit; shew that the perturbation in the mean distance of the planet is given by

$$\delta a = \frac{2Ma}{\mu}\left(1 + \tfrac{3}{4}\frac{b^2}{a^2}\right) e \cos\left(nt + \epsilon - \varpi\right),$$

M being the whole mass of the meteors, a, e, n, ϵ, ϖ being elements of the planet's orbit, and $b^3 : a^3$, $M^2 : \mu^2$, and e^2 being neglected.   (*M. T.* 1880.)

43. If a, b be the semiaxes of the instantaneous ellipse of a planet's orbit, and the only disturbing force be a tangential one whose acceleration is f, prove that

$$2a \frac{da}{dt} \frac{db}{dt} - b \left(\frac{da}{dt}\right)^2 = \frac{4f^2}{\mu} a^3 b.$$

(*M. T.* 1881.)

44. If H be proportional to the excess of the moment of momentum of the orbital motion of a disturbed planet about the Sun in a circular orbit at distance a above the value of the same thing in the actual elliptic orbit; then shew, from the equations for the variations of major-axis and excentricity, that, k being a constant, $\dfrac{dH}{dt} = -k \dfrac{dR}{d\varpi}$.
Also assuming that, when the inclination of the orbit to the plane of reference is zero, the variational equation for ϖ is

$$\frac{d\varpi}{dt} = \frac{na \sqrt{1-e^2}}{\mu e} \frac{dR}{de},$$

shew that
$$\frac{d\varpi}{dt} = k \frac{dR}{dH}.$$

(*M. T.* 1882.)

45. Explain how Elliptic Integrals and Bessel's Functions arise in the development of the disturbing function.

(*M. T.* 1879.)

46. If m, m' denote the masses of two planets, n, n', their mean motions, and a, a', their mean distances from the Sun, and if

$$m' \frac{a}{a'^2} C \cos \{(i'n' - in)\, t + \gamma\},$$

$$m \frac{a'}{a^2} C' \cos \{(i'n' - in)\, t + \gamma\},$$

represent any two corresponding terms in the respective developments of the portions

$$m' \frac{r}{r'^2} \cos (r, r') \quad \text{and} \quad m \frac{r'}{r^2} \cos (r, r'),$$

of the two disturbing functions, prove that

$$i^2 C = i'^2 C'.$$

(Smith's Prizes, 1880.)

47. Shew that the use of the ordinary disturbing function is attended with certain inconveniences in the case where we have to determine the perturbations of a superior planet produced by the action of an inferior one. Also shew that these inconveniences may be avoided by referring the motion of the superior planet to the common centre of gravity of the Sun and inferior planet as origin, and by taking a quantity proportional to $\dfrac{M}{R} + \dfrac{\mu}{\rho} - \dfrac{M+\mu}{r}$ as the disturbing function, where M is the Sun's mass, μ that of the disturbing planet, and R, ρ, and r are the distances of the disturbed planet from the Sun, from the disturbing planet, and from the centre of gravity of these two bodies respectively. How would you develope the above disturbing function?   (*S. P.* 1880.)

48. Considering the secular variations of the excentricities and perihelion longitudes of two mutually disturbing planets; prove that in the case where $m \sqrt{a} = m' \sqrt{a'}$, e and e' have the same greatest and least values E_1 and E_2, and $\varpi \sim \varpi'$ is never greater than $\tan^{-1} \left(\dfrac{E_1^2 - E_2^2}{2 E_1 E_2} \right)$.

(M. T. 1874.)

49. If there were only two planets moving about the Sun and at any instant the orbit of one were circular, the eccentricity of its orbit would always be given by

$$c \cdot \cos(\alpha t + \beta)$$

where c, α, β are constants, and the longitude of its perihelion would increase uniformly with the time.

(M. T. 1876.)

50. If γ be the angle between the orbits of two mutually disturbing planets, prove that approximately,

$$m \sqrt{a} \cdot e^2 + m' \sqrt{a'} \cdot e'^2 + \frac{4mm' \sqrt{aa'} \cdot \sin^2 \tfrac{1}{2}\gamma}{m \sqrt{a} + m' \sqrt{a'}}$$

is constant. (Pontécoulant, Système du Monde, Tome I. pp. 458—460.)

51. If l be the mean anomaly, v the true anomaly, g the angular distance of the perihelion from the ascending node, h the longitude of the ascending node in the fixed plane of reference,

$$L = \sqrt{\mu a}, \quad G = \sqrt{\mu a (1 - e^2)}, \quad H = \sqrt{\mu a (1 - e^2)} \cos i,$$

the other letters having their usual meanings, and if R denote the ordinary disturbing function plus the term $\dfrac{\mu^2}{2L^2}$, prove that x, y, z, the co-ordinates of the planet referred to the fixed plane as the plane of xy and to the line from which the longitudes are measured as the axis of x, are given by the equations

$$Gx = Gr \cos(v + g) \cos h - Hr \sin(v + g) \sin h,$$

$$Gy = Gr \cos(v + g) \sin h + Hr \sin(v + g) \cos h,$$

$$Gz = \sqrt{G^2 - H^2} \cdot r \sin(v + g),$$

and taking l, g, h, L, G, H, as the six elements, prove that

$$\frac{dL}{dt} = \frac{dR}{dl}, \quad \frac{dG}{dt} = \frac{dR}{dg}, \quad \frac{dH}{dt} = \frac{dR}{dh}.$$

(M. T. 1878.)

52. If M, m be the masses of Sun and disturbed planet, Ω the disturbing function due to the action of another planet; ϖ, ν the longitudes of the perihelion and ascending node, ϵ the time of passage through perihelion; a, e, i, the semiaxis major, excentricity and inclination of the orbit to a fixed plane; further if

$$\mu = -\frac{M + m}{2a},$$

$$\kappa = \sqrt{M + m} \sqrt{a(1 - e^2)},$$

$$\lambda = \sqrt{M + m} . \sqrt{a(1 - e^2)} (1 - \cos i):$$

then the pairs of elements (ϵ, μ), (κ, ϖ), (ν, λ) will be *canonical*, i.e.

$$\frac{d\epsilon}{dt} = \frac{d\Omega}{d\mu}, \quad \frac{d\mu}{dt} = -\frac{d\Omega}{d\epsilon}, \text{ \&c.}$$

(S. P. 1881.)

APPENDIX.

1. ON referring to Art. 39, it will be seen that the formulæ which have been obtained for calculating the elements of the orbit involve only partial differential coefficients of R with respect to these elements, multiplied by functions of the elements which do not contain the time explicitly. As it is to this circumstance that these formulæ mainly owe their advantage (since it renders them fit for use as soon as the partial differential coefficients have been calculated), it will be interesting to shew à priori that whatever system of elements be adopted, the formulæ for their calculation can always be arranged in this way.

2. If the motion of the planet be referred to three rectangular axes originating in the centre of gravity of the Sun, we have the equations of motion (see Art. 9),

$$\frac{d^2x}{dt^2} + \frac{\mu x}{r^3} = \frac{dR}{dx} \quad \dots\dots\dots\dots\dots\dots(1),$$

$$\frac{d^2y}{dt^2} + \frac{\mu y}{r^3} = \frac{dR}{dy} \quad \dots\dots\dots\dots\dots\dots(2),$$

$$\frac{d^2z}{dt^2} + \frac{\mu z}{r^3} = \frac{dR}{dz} \quad \dots\dots\dots\dots\dots\dots(3).$$

Let a, b, c, d, e, f be the six elements introduced by integrating these equations when $R = 0$, and for $\frac{dx}{dt}, \frac{dy}{dt}, \frac{dz}{dt}$ write x', y', z': then x', y' and z' can be expressed as functions of t and the elements; hence

$$\frac{dx'}{dt} = \left(\frac{dx'}{dt}\right) + \frac{dx'}{da}\frac{da}{dt} + \frac{dx'}{db}\frac{db}{dt} + \ldots\ldots,$$

where in $\left(\frac{dx'}{dt}\right)$ the elements are supposed constant.

If in equation (1) we put R equal to 0, we have

$$\left(\frac{dx'}{dt}\right) + \frac{\mu x}{r^3} = 0 ;$$

therefore $\qquad \dfrac{dx'}{dt} - \left(\dfrac{dx'}{dt}\right) = \dfrac{dR}{dx} ;$

therefore $\qquad \dfrac{dR}{dx} = \dfrac{dx'}{da}\dfrac{da}{dt} + \dfrac{dx'}{db}\dfrac{db}{dt} + \ldots\ldots,$

and similar equations hold for $\dfrac{dR}{dy}$ and $\dfrac{dR}{dz}$.

Now since R is a function of x, y and z,

$$\frac{dR}{da} = \frac{dR}{dx}\frac{dx}{da} + \frac{dR}{dy}\frac{dy}{da} + \frac{dR}{dz}\frac{dz}{da}$$

$$= \left(\frac{dx}{da}\frac{dx'}{da} + \frac{dy}{da}\frac{dy'}{da} + \frac{dz}{da}\frac{dz'}{da}\right)\frac{da}{dt}$$

$$+ \left(\frac{dx}{da}\frac{dx'}{db} + \frac{dy}{da}\frac{dy'}{db} + \frac{dz}{da}\frac{dz'}{db}\right)\frac{db}{dt}$$

$$+ \left(\frac{dx}{da}\frac{dx'}{dc} + \frac{dy}{da}\frac{dy'}{dc} + \frac{dz}{da}\frac{dz'}{dc}\right)\frac{dc}{dt}$$

$$+ \ldots\ldots\ldots\ldots\ldots\ldots\ldots\ldots$$

3. We may eliminate $\dfrac{da}{dt}$ from this expression: for, sup-posing x, y, and z expressed as functions of t and the elements, we have

$$\frac{dx}{dt} = \left(\frac{dx}{dt}\right) + \frac{dx}{da}\frac{da}{dt} + \frac{dx}{db}\frac{db}{dt} + \dots;$$

but by the principles of the method of the Variation of Parameters

$$\frac{dx}{dt} = \left(\frac{dx}{dt}\right);$$

therefore $\qquad \dfrac{dx}{da}\dfrac{da}{dt} + \dfrac{dx}{db}\dfrac{db}{dt} + \dfrac{dx}{dc}\dfrac{dc}{dt} + \dots = 0.$

Similarly, $\dfrac{dy}{da}\dfrac{da}{dt} + \dfrac{dy}{db}\dfrac{db}{dt} + \dfrac{dy}{dc}\dfrac{dc}{dt} + \dots = 0,$

$$\dfrac{dz}{da}\dfrac{da}{dt} + \dfrac{dz}{db}\dfrac{db}{dt} + \dfrac{dz}{dc}\dfrac{dc}{dt} + \dots = 0.$$

Multiplying these equations by $\dfrac{dx'}{da}$, $\dfrac{dy'}{da}$, &c., and adding, we obtain

$$0 = \left(\frac{dx}{da}\frac{dx'}{da} + \frac{dy}{da}\frac{dy'}{da} + \frac{dz}{da}\frac{dz'}{da}\right)\frac{da}{dt}$$

$$+ \left(\frac{dx}{db}\frac{dx'}{da} + \frac{dy}{db}\frac{dy'}{da} + \frac{dz}{db}\frac{dz'}{da}\right)\frac{db}{dt}$$

$$+ \left(\frac{dx}{dc}\frac{dx'}{da} + \frac{dy}{dc}\frac{dy'}{da} + \frac{dz}{dc}\frac{dz'}{da}\right)\frac{dc}{dt}$$

$$+ \dots\dots\dots\dots\dots$$

If this expression be subtracted from that for $\dfrac{dR}{da}$ in Art. 2, the latter may be written

$$\frac{dR}{da} = [a,\, b]\frac{db}{dt} + [a,\, c]\frac{dc}{dt} + \dots,$$

where $[a,\, b] = \dfrac{dx}{da}\dfrac{dx'}{db} - \dfrac{dx}{db}\dfrac{dx'}{da} + \dfrac{dy}{da}\dfrac{dy'}{db} - \dfrac{dy}{db}\dfrac{dy'}{da}$

$$+ \frac{dz}{da}\frac{dz'}{db} - \frac{dz}{db}\frac{dz'}{da}.$$

Similarly,

$$\frac{dR}{db} = [b,\, a]\frac{da}{dt} + [b,\, c]\frac{dc}{dt} + \dots,\ \&\text{c.}$$

4. By successive elimination between these equations, we can obtain expressions for $\dfrac{da}{dt}$, $\dfrac{db}{dt}$, &c., in terms of $\dfrac{dR}{da}$, $\dfrac{dR}{db}$, &c., $[a,\, b]$, $[a,\, c]$, &c.: if, then, we can shew that $[a,\, b]$, $[a,\, c]$, &c., are independent of the time explicitly, it will follow that this is also the case with the coefficients of $\dfrac{dR}{da}$, $\dfrac{dR}{db}$, &c., in the expressions for $\dfrac{da}{dt}$, $\dfrac{db}{dt}$, &c.

5. *To shew that* $[\mathrm{a},\, \mathrm{b}]$ *is independent of the time explicitly.*

Let $V = \dfrac{\mu}{r}$; then the equations of motion give

$$\left(\frac{dx'}{dt}\right) = \frac{dV}{dx}, \quad \left(\frac{dy'}{dt}\right) = \frac{dV}{dy}, \quad \left(\frac{dz'}{dt}\right) = \frac{dV}{dz}.$$

Now differentiating with respect to t only so far as it occurs explicitly,

$$\frac{d}{dt}[a,\, b] = \frac{dx}{da}\frac{d}{dt}\left(\frac{dx'}{db}\right) + \frac{dx'}{db}\frac{d}{dt}\left(\frac{dx}{da}\right)$$

$$- \frac{dx}{db}\frac{d}{dt}\left(\frac{dx'}{da}\right) - \frac{dx'}{da}\frac{d}{dt}\left(\frac{dx}{db}\right)$$

$$+ \dots\dots\dots\dots$$

$$= \frac{dx}{da}\frac{d}{db}\left(\frac{dx'}{dt}\right) + \frac{dx'}{db}\frac{dx'}{da}$$

$$- \frac{dx}{db}\frac{d}{da}\left(\frac{dx'}{dt}\right) - \frac{dx'}{da}\frac{dx'}{db}$$

$$+ \cdots\cdots\cdots$$

$$= \frac{dx}{da}\frac{d}{db}\left(\frac{dV}{dx}\right) - \frac{dx}{db}\frac{d}{da}\left(\frac{dV}{dx}\right)$$

$$+ \frac{dy}{da}\frac{d}{db}\left(\frac{dV}{dy}\right) - \frac{dy}{db}\frac{d}{da}\left(\frac{dV}{dy}\right)$$

$$+ \frac{dz}{da}\frac{d}{db}\left(\frac{dV}{dz}\right) - \frac{dz}{db}\frac{d}{da}\left(\frac{dV}{dz}\right)$$

$$= \frac{dx}{da}\frac{d}{dx}\left(\frac{dV}{db}\right) + \frac{dy}{da}\frac{d}{dy}\left(\frac{dV}{db}\right) + \frac{dz}{da}\frac{d}{dz}\left(\frac{dV}{db}\right)$$

$$- \frac{dx}{db}\frac{d}{dx}\left(\frac{dV}{da}\right) - \frac{dy}{db}\frac{d}{dy}\left(\frac{dV}{da}\right) - \frac{dz}{db}\frac{d}{dz}\left(\frac{dV}{da}\right)$$

$$= \frac{d^2V}{da\,db} - \frac{d^2V}{db\,da} = 0.$$

Hence [a, b] does not contain the time explicitly. The same is of course true of [a, c], [b, c], &c. It follows, then, that whatever system of elements be adopted, we can always express their differential coefficients in terms of the partial differential coefficients of R with respect to them, multiplied by functions of the elements which do not involve the time explicitly.

COROLLARY. We now see that in the expression for [a, b], viz. :—

$$\left(\frac{dx}{da}\frac{dx'}{db} - \frac{dx}{db}\frac{dx'}{da}\right) + \left(\frac{dy}{da}\frac{dy'}{db} - \frac{dy}{db}\frac{dy'}{da}\right) + \left(\frac{dz}{da}\frac{dz'}{db} - \frac{dz}{db}\frac{dz'}{da}\right),$$

we may replace x, y, z, x', y', z', by $x_0, y_0, z_0, x_0', y_0', z_0'$, their values at the origin of the time, without altering the value of this expression. We may next adopt instead of the constants a, b, c, d, e, f, the same quantities $x_0, y_0, z_0, \dot{x}_0', y_0', z_0'$. In this case we have evidently

$$[x_0, x_0'] = 1, \quad [y_0, y_0'] = 1, \quad [z_0, z_0'] = 1,$$
$$[x_0', x_0] = -1, \quad [y_0', y_0] = -1, \quad [z_0', z_0] = -1,$$

and all the analogous expressions different from these will vanish. The final equation of Art. 3 of this Appendix will give

$$\frac{dx_0}{dt} = \frac{dR}{dx_0'}, \qquad \frac{dy_0}{dt} = \frac{dR}{dy_0'}, \qquad \frac{dz_0}{dt} = \frac{dR}{dz_0'},$$

$$\frac{dx_0'}{dt} = -\frac{dR}{dx_0}, \qquad \frac{dy_0'}{dt} = -\frac{dR}{dy_0}, \qquad \frac{dz_0'}{dt} = -\frac{dR}{dz_0},$$

and no other combinations of the arbitrary constants will give simpler expressions than these for the variations of the constants adopted. From them the variations of the usual elliptic elements might be derived by proper analytical transformations. The group of constants x_0, y_0, z_0, x_0', y_0', z_0' may be termed *canonical elements*.

6. From the formula of Art. 3 of this Appendix, which is due to Lagrange, those of Chapter II. may be deduced: for this we refer to Pontécoulant's *Système du Monde*, Tome I. p. 542.

Thus, if we denote by u the excentric anomaly of the planet, by $nt + l$ its mean longitude at the time t, by e the excentricity of the orbit, and by a the semi-major axis, connected with the constant n by the equation $n^2 a^3 = \mu$, we shall have by the formulæ of elliptic motion

$$nt + l = u - e \sin u.$$

Next, if we denote by X, Y, the two rectangular co-ordinates of m referred to the plane of the orbit, the major-axis and latus-rectum being the co-ordinate axes, and the origin consequently at the focus of the curve, we shall have

$$X = a \cos u - ae, \qquad Y = a \sqrt{1 - e^2} \sin u.$$

Lastly, if we denote by ν the angle which the major-axis of the orbit makes with the *node* or line of intersection of the plane of this orbit and the fixed plane of xy, by Ω the angle which the node makes with the axis of x, and by i the inclination of the orbit to the plane of xy; we shall then have by transformation of co-ordinates,

$$x = (X\cos\nu - Y\sin\nu)\cos\Omega - (X\sin\nu + Y\cos\nu)\cos i \,.\, \sin\Omega,$$
$$y = (X\cos\nu - Y\sin\nu)\sin\Omega + (X\sin\nu + Y\cos\nu)\cos i \,.\, \cos\Omega,$$
$$z = (X\sin\nu + Y\cos\nu)\sin i.$$

These values of x, y, z are very convenient for use in determining the values of the expressions $[a, b]$, $[a, c]$, $[b, c]$, &c., since the constants in the expressions for x, y, z, are separated into two groups; the variables X, Y only containing a, e, l, which depend on the form of the orbit and on the position of the planet on that curve at a given instant, being independent of the three constants ν, Ω, i, which fix the position of the major-axis and of the plane of the orbit.

By the aid of these values and of their partial differential coefficients substituted in forms such as

$$\left(\frac{dx}{da}\frac{dx'}{db} - \frac{dx}{db}\frac{dx'}{da}\right) + \left(\frac{dy}{da}\frac{dy'}{db} - \frac{dy}{db}\frac{dy'}{da}\right) + \left(\frac{dz}{da}\frac{dz'}{db} - \frac{dz}{db}\frac{dz'}{da}\right)$$

we shall determine the values of the fifteen symbols $[a, l]$, $[a, e]$, &c., and by an easy calculation obtain the following values:

$$[a,\, l] = -\frac{an}{2}, \quad [a,\, \nu] = -\frac{\sqrt{a\,(1 - e^2)}}{2a}, \quad [e,\, \nu] = \frac{ae}{\sqrt{a\,(1 - e^2)}},$$

$$[a,\, \Omega] = \frac{\cos i\,\sqrt{a\,(1 - e^2)}}{2a}, \quad [e,\, \Omega] = -\frac{ae\cos i}{\sqrt{a\,(1 - e^2)}},$$

$$[i,\, \Omega] = \sin\phi\,\sqrt{a\,(1 - e^2)}.$$

The values of the nine other symbols $[a, e]$, $[a, i]$, $[e, l]$, $[e, \Omega]$, $[l, i]$, $[l, \Omega]$, $[\Omega, \nu]$, $[l, \nu]$, $[i, \nu]$ are equal to zero.

After substituting these values in the general formula at the end of Art. 3 of this Appendix we shall find

$$\frac{dR}{da} = -\frac{an}{2}\frac{dl}{dt} - \frac{\sqrt{a(1-e^2)}}{2a}\frac{dv}{dt} - \frac{\cos i \sqrt{a(1-e^2)}}{2a}\frac{d\Omega}{dt},$$

$$\frac{dR}{de} = \frac{ae}{\sqrt{a(1-e^2)}}\frac{dv}{dt} + \frac{ae\cos i}{\sqrt{a(1-e^2)}}\frac{d\Omega}{dt},$$

$$\frac{dR}{dl} = \frac{an}{2}\frac{da}{dt},$$

$$\frac{dR}{dv} = \frac{\sqrt{a(1-e^2)}}{2a}\frac{da}{dt} - \frac{ae}{\sqrt{a(1-e^2)}}\frac{de}{dt},$$

$$\frac{dR}{d\Omega} = \frac{\cos i \sqrt{a(1-e^2)}}{2a}\frac{da}{dt} - \frac{ae\cos i}{\sqrt{a(1-e^2)}}\frac{de}{dt} - \sin i \sqrt{a(1-e^2)}\frac{di}{dt},$$

$$\frac{dR}{di} = \sin i \sqrt{a(1-e^2)}\frac{d\Omega}{dt}.$$

Whence, by an easy elimination, and observing that $n^2a^3 = \mu$, we deduce

$$\frac{da}{dt} = \frac{2na^2}{\mu}\frac{dR}{dl},$$

$$\frac{de}{dt} = \frac{an(1-e^2)}{e\mu}\frac{dR}{dl} - \frac{an\sqrt{(1-e^2)}}{e\mu}\frac{dR}{dv},$$

$$\frac{dl}{dt} = -\frac{2a^2n}{\mu}\frac{dR}{da} - \frac{an(1-e^2)}{\mu e}\frac{dR}{de},$$

$$\frac{dv}{dt} = \frac{an\sqrt{1-e^2}}{e\mu}\frac{dR}{de} - \frac{an\cos i}{\mu\sin i\sqrt{(1-e^2)}}\frac{dR}{di},$$

$$\frac{d\Omega}{dt} = \frac{an}{\mu\sin i\sqrt{(1-e^2)}}\frac{dR}{di},$$

$$\frac{di}{dt} = \frac{an\cos i}{\mu\sin i\sqrt{(1-e^2)}}\frac{dR}{dv} - \frac{an}{\mu\sin i\sqrt{(1-e^2)}}\frac{dR}{d\Omega};$$

formulæ from which we can readily obtain those of Art. 39 of this work when we take into consideration that

$$l = \epsilon - \varpi, \quad v = \varpi - \Omega.$$

For by Arts. 11 and 13, R may be expressed as

$$\phi\,(a,\,e,\,i,\,r,\,\theta - \Omega,\,\Omega),$$

or as either of $\phi\,(a,\,e,\,i,\,nt+\epsilon-\varpi,\,\varpi-\Omega,\,\Omega)\equiv R_2$,

$$\phi\,(a,\,e,\,i,\,nt+l,\,\nu,\,\Omega)\equiv R_1,$$

and consequently

$$\frac{dR_2}{d\varpi}=-\frac{dR_1}{dl}+\frac{dR_1}{d\nu},\quad\frac{dR_2}{d\epsilon}=\frac{dR_1}{dl},$$

$$\frac{dR_2}{d\Omega}=\frac{dR_1}{d\Omega}-\frac{dR_1}{d\nu},$$

and so

$$\frac{dR_1}{dl}=\frac{dR_2}{d\epsilon},\quad\frac{dR_1}{d\nu}=\frac{dR_2}{d\varpi}+\frac{dR_2}{d\epsilon},$$

$$\frac{dR_1}{d\Omega}=\frac{dR_2}{d\Omega}+\frac{dR_2}{d\varpi}+\frac{dR_2}{d\epsilon},$$

by aid of which the needful changes may be effected.

NOTES TO ARTS. 31 AND 33.

7.   The process of Arts. 31 and 33 may perhaps present some difficulty to the student, arising from the fact that the method of determining the position of the planet by its polar co-ordinates on the plane of the orbit, i.e. on a plane moving with the planet, is not applicable to the geometrical purpose of defining the position of a point on the curve of reference, a curve which, although it passes through the planet at some given instant, has nothing to do with the planet's subsequent motion.

Of course the position of any point may be determined by its polar co-ordinates on a plane passing through it and some fixed point or origin, the inclination of this plane to some fixed plane through the origin, and the longitude of its node.   But inasmuch as an infinite number of planes can be drawn through two given points, some further condition is necessary to regulate the motion of that on which the polar co-ordinates are measured. If we are concerned with the motion of the planet, the condition is that this plane shall contain the direction of the planet's motion at the instant under consideration, in which case it becomes the plane of the orbit: if we are concerned with any other point, then some different condition is requisite. In applying this method to the curve of reference, we are, of course, at liberty to assign any condition we please, but it will be convenient that it be such that the co-ordinates of that point of the curve of reference which coincides with the planet shall be identical with those of the planet itself: this object is attained by the two conditions imposed in Arts. 31 and 33.

ON THE EQUATIONS OF MOTION OF A DISTURBED PLANET.

8. The principles of the *conservation of areas* and of the *vis viva* being applicable to the motion of the Sun and any number of mutually disturbing planets, four first integrals of the equations of motion may be found. Although these integrals are of little use in determining the motions of the several bodies, which cannot be completely found except by methods of approximation, we shall shew how they may be obtained in the case of the Sun and two planets.

The differential equations of motion of the two planets relative to the Sun are

$$\left.\begin{aligned}
\frac{d^2x}{dt^2} &= -\frac{M+m}{r^3}\,x + \frac{x'-x}{\rho^3}\,m' - \frac{m'x'}{r'^3}\\[4pt]
\frac{d^2y}{dt^2} &= -\frac{M+m}{r^3}\,y + \frac{y'-y}{\rho^3}\,m' - \frac{m'y'}{r'^3}\\[4pt]
\frac{d^2z}{dt^2} &= -\frac{M+m}{r^3}\,z + \frac{z'-z}{\rho^3}\,m' - \frac{m'z'}{r'^3}
\end{aligned}\right\}\dots\dots(1),$$

$$\left.\begin{aligned}
\frac{d^2x'}{dt^2} &= -\frac{M+m'}{r'^3}\,x' + \frac{x-x'}{\rho^3}\,m - \frac{mx}{r^3}\\[4pt]
\frac{d^2y'}{dt^2} &= -\frac{M+m'}{r'^3}\,y' + \frac{y-y'}{\rho^3}\,m - \frac{my}{r^3}\\[4pt]
\frac{d^2z'}{dt^2} &= -\frac{M+m'}{r'^3}\,z' + \frac{z-z'}{\rho^3}\,m - \frac{mz}{r^3}
\end{aligned}\right\}\dots\dots(2),$$

where $\rho = \{(x'-x)^2 + (y'-y)^2 + (z'-z)^2\}^{\frac{1}{2}}.$

From these we find

$$m\left(y\frac{d^2z}{dt^2} - z\frac{d^2y}{dt^2}\right) + m'\left(y'\frac{d^2z'}{dt^2} - z'\frac{d^2y'}{dt^2}\right)$$

$$= mm'\left(\frac{1}{r^3} - \frac{1}{r'^3}\right)(yz' - y'z),$$

and

$$(my + m'y')\left(m\frac{d^2z}{dt^2} + m'\frac{d^2z'}{dt^2}\right) - (mz + m'z')\left(m\frac{d^2y}{dt^2} + m'\frac{d^2y'}{dt^2}\right)$$

$$= (M + m + m')\,mm'\left(\frac{1}{r^3} - \frac{1}{r'^3}\right)(yz' - y'z).$$

Hence

$$m\left(y\frac{d^2z}{dt^2} - z\frac{d^2y}{dt^2}\right) + m'\left(y'\frac{d^2z'}{dt^2} - z'\frac{d^2y'}{dt^2}\right)$$

$$= \frac{my + m'y'}{M+m+m'}\left(m\frac{d^2z}{dt^2} + m'\frac{d^2z'}{dt^2}\right) - \frac{mz + m'z'}{M+m+m'}\left(m\frac{d^2y}{dt^2} + m'\frac{d^2y'}{dt^2}\right).$$

Integrating, we have

$$m\left(y\frac{dz}{dt} - z\frac{dy}{dt}\right) + m'\left(y'\frac{dz'}{dt} - z'\frac{dy'}{dt}\right) + C_1$$

$$= \frac{my + m'y'}{M+m+m'}\left(m\frac{dz}{dt} + m'\frac{dz'}{dt}\right) - \frac{mz + m'z'}{M+m+m'}\left(m\frac{dy}{dt} + m'\frac{dy'}{dt}\right),$$

with two similar integrals.

These equations may be written thus

$$M\left\{m\left(y\frac{dz}{dt} - z\frac{dy}{dt}\right) + m'\left(y'\frac{dz'}{dt} - z'\frac{dy'}{dt}\right)\right\}$$

$$+ mm'\left\{(y'-y)\frac{d}{dt}(z'-z) - (z'-z)\frac{d}{dt}(y'-y)\right\} + C_1 = 0,$$

with two similar equations.

For a first approximation, in consequence of the magnitude of the Sun's mass by comparison with that of any of the planets, we may neglect the term multiplied by mm', and we have

$$m\left(y\frac{dz}{dt}-z\frac{dy}{dt}\right)+m'\left(y'\frac{dz'}{dt}-z'\frac{dy'}{dt}\right)=c_1$$

$$m\left(z\frac{dx}{dt}-x\frac{dz}{dt}\right)+m'\left(z'\frac{dx'}{dt}-x'\frac{dz'}{dt}\right)=c_2$$

$$m\left(x\frac{dy}{dt}-y\frac{dx}{dt}\right)+m'\left(x'\frac{dy'}{dt}-y'\frac{dx'}{dt}\right)=c_3$$

Now

$$\left(x\frac{dy}{dt}-y\frac{dx}{dt}\right)$$

is the projection of the area

$$\sqrt{(M+m)\,a\,(1-e^2)},$$

described in the instantaneous elliptic orbit, on the plane of xy, to which that orbit is inclined at an angle i. Hence

$$\left(x\frac{dy}{dt}-y\frac{dx}{dt}\right)=\sqrt{(M+m)\,a\,(1-e^2)}\,(1+\tan^2 i)^{-\frac{1}{2}},$$

and we have

$$m\sqrt{(M+m)\,a\,(1-e^2)}\,(1+\tan^2 i)^{-\frac{1}{2}}$$
$$+\,m'\sqrt{(M+m')\,a'\,(1-e'^2)}\,(1+\tan^2 i')^{-\frac{1}{2}}=c_3,$$

or, retaining only terms affected by M,

$$m\sqrt{a\,(1-e^2)}\,(1+\tan^2 i)^{-\frac{1}{2}}$$
$$+\,m'\sqrt{a'\,(1-e'^2)}\,(1+\tan^2 i')^{-\frac{1}{2}}=\text{constant.}$$

If we multiply the differential equations (1) of motion of the planet m by the factors

$$2m\frac{dx}{dt}-2m\cdot\frac{m\dfrac{dx}{dt}+m'\dfrac{dx'}{dt}}{M+m+m'},$$

$$2m\frac{dy}{dt}-2m\cdot\frac{m\dfrac{dy}{dt}+m'\dfrac{dy'}{dt}}{M+m+m'},$$

$$2m\frac{dz}{dt}-2m\cdot\frac{m\dfrac{dz}{dt}+m'\dfrac{dz'}{dt}}{M+m+m'},$$

and multiply also the differential equations (2) of motion of
the planet m' by similar factors, and add the six products, we
shall find

$$2m\left(\frac{dx}{dt}\frac{d^2x}{dt^2}+\frac{dy}{dt}\frac{d^2y}{dt^2}+\frac{dz}{dt}\frac{d^2z}{dt^2}\right)$$

$$+2m'\left(\frac{dz'}{dt}\frac{d^2x'}{dt^2}+\frac{dy'}{dt}\frac{d^2y'}{dt^2}+\frac{dz'}{dt}\frac{d^2z'}{dt^2}\right)$$

$$-2\frac{m\dfrac{dx}{dt}+m'\dfrac{dx'}{dt}}{M+m+m'}\left(m\frac{d^2x}{dt^2}+m'\frac{d^2x'}{dt^2}\right)$$

$$-2\frac{m\dfrac{dy}{dt}+m'\dfrac{dy'}{dt}}{M+m+m'}\left(m\frac{d^2y}{dt^2}+m'\frac{d^2y'}{dt^2}\right)$$

$$-2\frac{m\dfrac{dz}{dt}+m'\dfrac{dz'}{dt}}{M+m+m'}\left(m\frac{d^2z}{dt^2}+m'\frac{d^2z'}{dt^2}\right)$$

$$-2M.\left(\frac{m}{r^2}\frac{dr}{dt}+\frac{m'}{r'^2}\frac{dr'}{dt}\right)$$

$$-2\frac{d}{dt}\left(\frac{mm'}{\sqrt{(x'-x)^2+(y'-y)^2+(z'-z)^2}}\right)=0$$

or integrating

$$m\left\{\left(\frac{dx}{dt}\right)^2+\left(\frac{dy}{dt}\right)^2+\left(\frac{dz}{dt}\right)^2\right\}+m'\left\{\left(\frac{dx'}{dt}\right)^2+\left(\frac{dy'}{dt}\right)^2+\left(\frac{dz'}{dt}\right)^2\right\}$$

$$-\frac{\left\{\left(m\dfrac{dx}{dt}+m'\dfrac{dx'}{dt}\right)^2+\left(m\dfrac{dy}{dt}+m'\dfrac{dy'}{dt}\right)^2+\left(m\dfrac{dz}{dt}+m'\dfrac{dz'}{dt}\right)^2\right\}}{M+m+m'}$$

$$-2M\left(\frac{m}{r}+\frac{m'}{r'}\right)-\frac{2mm'}{\sqrt{(x'-x)^2+(y'-y)^2+(z-z)^2}}=\text{const.}$$

an equation which may be written

$$Mm\left\{\left(\frac{dx}{dt}\right)^2 + \left(\frac{dy}{dt}\right)^2 + \left(\frac{dz}{dt}\right)^2\right\} + Mm'\left\{\left(\frac{dx'}{dt}\right)^2 + \left(\frac{dy'}{dt}\right)^2 + \left(\frac{dz'}{dt}\right)^2\right\}$$

$$+ mm'\left\{\left(\frac{dx'}{dt} - \frac{dx}{dt}\right)^2 + \left(\frac{dy'}{dt} - \frac{dy}{dt}\right)^2 + \left(\frac{dz'}{dt} - \frac{dz}{dt}\right)^2\right\}$$

$$- 2(M+m+m')\left\{M\left(\frac{m}{r} + \frac{m'}{r'}\right) + \frac{mm'}{\sqrt{(x'-x)^2 + (y'-y)^2 + (z'-z)^2}}\right\}$$

$$= \text{const.},$$

an equation containing the principle of the *vis viva*. Pontécoulant, Tome I., p. 210.

EXAMPLE OF NUMERICAL CALCULATION.

9. By way of illustration, we will here shew how to calculate the inequality in the earth's radius vector due to the term

$$m'C_2 \cos 2\left\{(n-n')t + \epsilon - \epsilon'\right\}$$

in R, when Jupiter is the disturbing planet.

Following the method of Chapter VII., and taking the result of the first approximation, we have (Art. 110)

$$\delta r = \frac{P_1}{a\left\{n^2 - (2n-2n')^2\right\}} \cos 2\left\{(n-n')t + \epsilon - \epsilon'\right\},$$

or, writing for P_1 its value, the coefficient of this term becomes

$$- \frac{m'}{(3n-2n')(n-2n')}\left\{\frac{dC_2}{da} + \frac{2nC_2}{a(n-n')}\right\},$$

or since $n^2a^3 = \mu = M + m$, if M denote the Sun's mass, we may write, to the first order of the disturbing force,

$$- \frac{m'}{M}\frac{n^2}{(3n-2n')(n-2n')}\left\{a^3\frac{dC_2}{da} + \frac{2na^2C_2}{n-n'}\right\},$$

a more convenient expression; for since n occurs in the form of a ratio, it is no longer necessary to suppose it expressed in circular measure.

The values of C_2 and $\dfrac{dC_2}{da}$ must, of course, be found by the methods of Chapter II.: by slightly changing our notation, however, we shall be able to avail ourselves of the calculation of M. Bouvard, as given in the third volume of the *Mécanique Céleste*, and the third volume of Pontécoulant's *Théorie Analytique du Système du Monde.*

If $c_2 = a' C_2$, $\alpha = \dfrac{a}{a'}$, we have

$$C_2 = \frac{c_2}{a'}, \quad \frac{dC_2}{da} = \frac{1}{a'}\frac{dc_2}{da} = \frac{1}{a'^2}\frac{dc_2}{d\alpha} :$$

thus the coefficient becomes

$$-\frac{m'}{M}\frac{n^2}{(3n-2n')(n-2n')}\left\{ a\alpha^2\frac{dc_2}{d\alpha} + \frac{2n\alpha\alpha c_2}{n-n'}\right\},$$

or, taking the Sun's mass and the Earth's mean distance as the units of mass and of distance,

$$-\frac{m'n^2\alpha}{(3n-2n')(n-2n')}\left\{ \alpha\frac{dc_2}{d\alpha} + \frac{2nc_2}{n-n'}\right\}.$$

We take from the *Système du Monde*, Tome III.*,

$$m' = \frac{1}{1053\cdot924}, \quad n = 1295977''\cdot35, \quad n' = 109256''\cdot29,$$

$$\log_{10}\alpha = \bar{1}\cdot2838993, \quad \log_{10}c_2 = \bar{2}\cdot4501721,$$

$$\log_{10}\alpha\frac{dc_2}{d\alpha} = \bar{2}\cdot7581084 ;$$

* In the second edition of the first volume, published in 1856, more recent values of m', n and n' are given, but our object being simply to illustrate the method of calculation, the above values have been retained.

and with the aid of a table of logarithms the numerical value of the coefficient may easily be found. Thus, in consequence of the term in R we have been considering, we find in δr the term

$$- \cdot 0000092122 \cos 2 \left\{ (n - n') t + \epsilon - \epsilon' \right\}.$$

ON THE METHODS OF CALCULATING THE MASSES OF
THE PLANETS.

10.   There are in general two methods of determining the masses of the planets; either by observations on a satellite, when the planet is accompanied by a satellite; or by comparing the inequalities produced in their motion by their mutual action, as deduced from observation, with the same inequalities calculated from theory. The secular variations are best adapted to give the most exact results; but these are not yet known with sufficient accuracy to allow of this use. We are therefore obliged to recur to the periodic variations, and, by combining a vast number of observations, gather from them the most probable results[*].

11.   When the planet is accompanied by a satellite the formula for calculating its mass may be obtained as follows :

Let M, m, m' be the masses of the Sun, the planet, and the satellite: P, P' the periodic times of the planet about the Sun, and the satellite about the planet; a, a' the mean distances of the planet from the Sun, and the satellite from the planet. Then we have

$$P = \frac{2\pi a^{\frac{3}{2}}}{\sqrt{(M + m)}}, \quad P' = \frac{2\pi a'^{\frac{3}{2}}}{\sqrt{(m + m')}} ;$$

therefore $$\frac{m + m'}{M + m} = \frac{P^2 a'^3}{P'^2 a^3},$$

---

[*] Pontécoulant' *Système du Monde*, Tome III. p. 340.

or approximately,

$$\frac{m + m'}{M} = \frac{P^2 a'^3}{P'^2 a^3}.$$

This equation gives the mass of the planet when that of its satellite is known. If the latter be neglected, the formula becomes

$$\frac{m}{M} = \frac{P^2 a'^3}{P'^2 a^3}.$$

12. In the case of the Earth, this method is not sufficiently exact, but the following may be employed. The attraction of the Earth on a body at its surface, in the parallel of which the square of the sine of the latitude is $\frac{1}{3}$, is very nearly the same as if the Earth were condensed into its centre. (See Pratt's *Figure of the Earth*, Art. 89.) Let then $\sin^2 l = \frac{1}{3}$, $g =$ the Earth's attraction on a body at its surface in latitude l, b the mean radius of the Earth, E the mass of the Earth, M the mass of the Sun, P the length of the year, and a the mean radius of the Earth's orbit. Then

$$g = \frac{E}{b^2}, \qquad P = \frac{2\pi a^{\frac{3}{2}}}{\sqrt{M}};$$

therefore

$$\frac{E}{M} = \frac{g b^2 P^2}{4\pi^2 a^3} = \frac{g P^2}{4\pi^2 b} \left(\frac{b}{a}\right)^3,$$

where

$$\frac{b}{a} = \text{sine of Sun's parallax.}$$

For the methods of calculating the mass of the Moon we refer to Pontécoulant's *Système du Monde*, Tome IV. p. 651.

ON THE CONSTRUCTION OF ASTRONOMICAL TABLES.

13. In the present state of Astronomy, the values of
the masses and elements, upon which these tables depend,
may be considered as approximately known, so that small
corrections only are necessary: we propose, in this note, to
shew briefly how these corrections are effected.

A series of observations are taken in R.A. and N.P.D.,
separated by considerable intervals of time: from each of
these the geocentric longitude and latitude are obtained, and
thence the heliocentric longitude and latitude. These co-
ordinates are also calculated for each period of observation
by the methods of the preceding Chapters, employing the
existing values of the masses and elements. The differences
between the observed and calculated values will be due to
errors in the latter, for we assume that our observations can
be depended upon. From these errors in longitude and
latitude, the corresponding corrections in the values of the
masses and elements are obtained as follows:

Let θ denote the longitude of the planet as calculated
from theory, $\theta + \delta\theta$ its value as given by observation, m', m'',
&c., the existing values of the masses of the disturbing
planets, $\delta m', \delta m''$, &c., $\delta n, \delta e, \delta \varpi, \delta \epsilon$ the requisite corrections
to the values of the masses and elements. Then, retaining
only the first power of the excentricity and of the disturbing
force, we have by the methods of the preceding Chapters,

$$\theta = nt + \epsilon + 2e \sin (nt + \epsilon - \varpi) + m'P' + m''P'' + \ldots,$$

where $m'P', m''P''$, &c., are the terms due to the perturbations of the planets m', m'', &c. It follows that

$$\delta\theta = \frac{d\theta}{dn} \delta n + \frac{d\theta}{de} \delta e + \frac{d\theta}{d\varpi} \delta \varpi + \frac{d\theta}{d\epsilon} \delta \epsilon + P'\delta m' + P''\delta m'' + \ldots,$$

where $\delta\theta$, being the difference between the observed and calculated values of the longitude, is known.

A similar equation will result from each observation of the series, and the number of these must be reduced by the method of least squares to that of the unknown quantities: thus, by the solution of these equations, the required corrections to m', m'', &c., n, e, ϖ, ϵ will be obtained.

In like manner, from a comparison between the observed and calculated values of the latitude, the corrections to ι and Ω may be obtained.

Tables of the numerical values of the masses and elements will be found in Herschel's *Outlines of Astronomy*, and in the first volume of Pontécoulant's *Système du Monde*.

THE END.

CAMBRIDGE: PRINTED BY C. J. CLAY, M.A. & SON AT THE UNIVERSITY PRESS.

www.ingramcontent.com/pod-product-compliance
Lightning Source LLC
Chambersburg PA
CBHW021804190326
41518CB00007B/436